W9-BIV-312

101 PRACTICAL USES FOR
PROPANE TORCHES

BY ROBERT BRIGHTMAN

Illustrated by Henry Clark

TAB BOOKS
Blue Ridge Summit, Pa. 17214

FIRST EDITION

FIRST PRINTING—FEBRUARY 1978

Published by arrangement with Dorison House Publishers Inc.

Copyright © 1977 by Dorison House Publishers Inc.

Printed in the United States
of America

Library of Congress Cataloging in Publication Data

Brightman, Robert.
  101 practical uses for propane torches.

  1. Solder and soldering. 2. Brazing. 3. Welding.
I. Title. II. Title: Propane torches.
TT267.B69   621.9   78-2788
ISBN 0-8306-9970-7
ISBN 0-8306-1030-8 pbk.

Cover photo of Oxygen Cutting Torch courtesy of Bernzomatic Corporation Rochester, New York.

Contents

Caution

Whenever using any type of brazing or welding equipment always wear appropriate goggles and keep some sort of fire-fighting apparatus on hand, even if only a pail of water.

Introduction

According to U.S. Department of Commerce figures, American homeowners spent more than $9 billion on home repairs and improvements last year. But do-it-yourself handymen only spent about $3 billion. In other words, by doing many household chores, maintenance and repair jobs themselves, and not farming them out to professional tradesmen, the do-it-yourselfers were able to save *one out of every three dollars* spent for home repairs, maintenance and improvement.

You, too, can make these same substantial savings with the aid of common, everyday tools found in the home. Bear in mind that the chief difference between you and a professional plumber, welder or electrician is that he (or she!) has experience in doing a particular job and that experience means that he can do the job much faster than you can. But you can do just as good a job as the professional—and possibly even better. It will take you more time—your time—something you will not have to pay for on an hourly basis.

So, whether you own your home, rent or live in an apartment, this book will help you solve many of those annoying problems that always seem to crop up and plague us. Most repair jobs are usually within the capabilities of the average man or woman. All they need is a little help, which this book will provide. You will find this book especially handy when any job requiring the application of heat is called for—laying and removing tiles, repairing roof gutters, arts and crafts work, soldering, welding, brazing and cutting metals. Check the index for the topic you are interested in.

Dedicated to my wife, Mollie, who proofread my prose, sutured many split infinitives and snipped some dangling participles.

Robert Brightman

Chapter 1

Soldering

Soldering is the joining of two pieces of metal by means of a low-melting alloy called solder. This solder always has a lower melting temperature than the metals to be joined. Soldering is not welding, as the metals to be joined are not melted in the process—only the solder.

Soldering falls into 2 types—soft soldering and hard soldering, better known as "brazing." Soft solders melt at comparatively low temperatures, below 700°F, and are the type of soldering generally used in the home workshop. These so-called "soft" solders make good joints between metals such as copper, brass, tin and steel. Soft solders are made of lead and tin with the proportion of each varying between 48% and 52%—usually called "half-and-half." This type of solder melts at 370°F, a temperature low enough to join the metals mentioned above.

Hard solders are alloys of gold, silver, copper, lead, tin and bismuth. These solders melt at much higher temperatures than the soft solders. Care should be used when working with hard solders as often their melting points will be quite near those of the metals—in thin stock—to be joined.

Regardless of the type of solder that will be used—hard or soft—some kind of soldering flux is most important. The flux prevents an oxide from forming on the metals to be joined. An oxide on the metal will prevent successful soldering. An acid-type flux is generally used for heavy-duty soldering such as when joining sheet metal and roof repair work. A rosin-type solder is invariably

The first step in any soldering operation is to "tin" the iron. Clean it with steel wool, dip in flux and apply a thin coat of solder to iron.

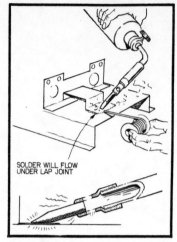

SOLDER WILL FLOW UNDER LAP JOINT

Small lap joint can be soldered with the soldering tip, the flame is not required. Wipe off excess flux while the joint is still warm.

used when soldering involves electrical equipment. Fluxes for hard soldering are especially formulated for the particular metal they are to be used on.

Soft Soldering

A cardinal rule to be observed when doing any kind of soldering work is cleanliness. We know of a professional mechanic who will actually wash his hands with soap and water before starting *any* soldering.

The first step is to clean the soldering tip on your propane gas torch. Use steel wool or sandpaper. Wipe off any grit with a clean cloth. Mount the soldering tip on the *pencil burner only*. Crack the valve and light with a match or the sparker. As soon as the tip is hot enough to melt solder, about 30 seconds or so, dip the tip in a flux paste or sal-ammoniac and apply the solder to the tip until it is completely covered with solder. Wipe off any excess solder with a clean rag leaving the tip bright and shiny with solder. Keep the flame small at all times. A large flame will overheat the tip requiring retinning, as this operation is called.

The soldering tip is used on smaller jobs and electrical connections. For larger jobs, such as drainpipes, use the pencil flame burner on your propane torch.

Now apply your attention to the work to be soldered. Thoroughly clean all the surfaces to be soldered with sandpaper or steel wool and apply a thin coating of flux. Heat the surfaces to

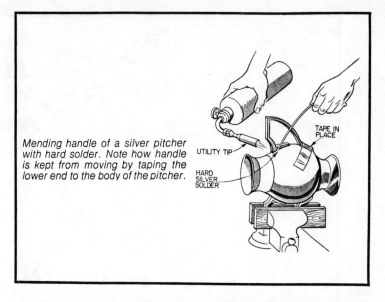

Mending handle of a silver pitcher with hard solder. Note how handle is kept from moving by taping the lower end to the body of the pitcher.

be soldered with the soldering tip. Keep the face of the tip flat against the surfaces to be soldered. This will serve to preheat the metal. The next step is to apply the solder to the surfaces to be soldered. Apply the solder to the work—not to the soldering tip.

Brazing or Hard Soldering

Hard soldering on such metals as silver, gold, steel and bronze makes a much stronger—and neater—joint than soft soldering. A propane torch will not produce sufficient heat to efficiently braze, or hard solder. Higher heat brazing torches, such as the BernzOmatic Super Torch, are recommended. The Super Torch operates just like a propane torch. In fact, the higher heat output of the Super Torch will allow faster soft soldering as well as brazing.

As in soft soldering, the first step is to carefully clean the area where the pieces are to be joined. Sandpaper, emery cloth, a file or steel wool can be used for this step. If at all possible, clamp the pieces to be joined so that they will not move during the operation. Apply the proper flux to the metal—type of flux depends upon the solder being used and the metal to be joined. There are more than a dozen different types of brazing rods and fluxes.

When working with brazing rods, use a rod that matches the color of the work to be joined. If you are joining silver, for example, use a silver solder. This is an alloy of silver (8 parts),

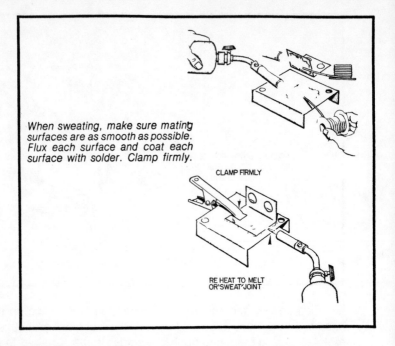

When sweating, make sure mating surfaces are as smooth as possible. Flux each surface and coat each surface with solder. Clamp firmly.

CLAMP FIRMLY

RE HEAT TO MELT OR 'SWEAT' JOINT

copper (3 parts) and zinc (1 part). If the job involves soldering gold, use a rod containing gold, silver and copper.

An inexpensive flux for fairly large jobs consists of powdered borax mixed with water to the consistency of cream. Heat the joint with the torch until the solder starts to melt when it is applied to the metal. Use the tip of the inner core of the flame which will be the hottest part.

Many brazing rods are flux coated or flux cored. These require no additional flux.

The melted flux left behind after the job is finished will leave a dark brown residue. It can be removed from the soldered joint by immersing the item in a solution of 1 part sulphuric acid to 2 parts water. *Important: Always add acid to water, never water to acid.* Best to wear rubber gloves and goggles when doing this job. Allow the work to soak in this solution for about 20 minutes. Flux can also be removed by immersion in boiling water, provided this is done before the joint has cooled.

Sweating

Sweating is another form of soldering for uniting the work. In this operation it is most important that the mating surfaces of the items be as smooth as possible and make a good mechanical fit

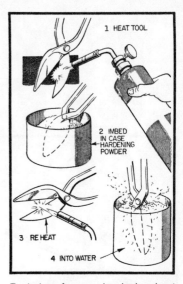

Large lap joints should be soldered with the torch flame. Solder, because of capillary action, flows to the hottest part of the joint.

Technique for case-hardening; heat the tool, imbed in the special powder, reheat the tool, and then immerse in water for the final step.

to each other. As in all types of soldering, thoroughly clean the surfaces to be joined. Steel wool is good for this purpose. Flux both surfaces and coat each surface with solder—this is a form of tinning. Assemble the 2 pieces and if at all possible clamp them together so they will have no chance of moving during the sweating process.

The next step is to apply the propane gas torch until the solder *within* the joint remelts and unites the 2 pieces. It is imperative that the joints be held in position until the solder is cool. Additional solder can be added, for the sake of strength, around the outside of the joint forming a fillet of metal.

Any flux that has oozed out can be wiped away with a rag while the metal is still *warm*. But don't do any wiping while the metal is *hot*, otherwise you may disturb the solder and loosen the joint.

Sheet-Metal Work
Whether you're applying a patch to a gutter or making a copper planter, it's important to remember that solder, itself, is not a strong material. You can't make a good joint simply by bringing 2 metal edges together and running a ribbon of solder over the seam. Instead, use either a lap joint or, when an assembly will

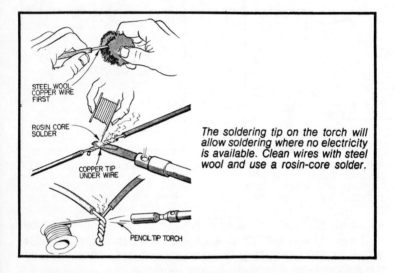

STEEL WOOL COPPER WIRE FIRST

ROSIN CORE SOLDER

COPPER TIP UNDER WIRE

PENCIL TIP TORCH

The soldering tip on the torch will allow soldering where no electricity is available. Clean wires with steel wool and use a rosin-core solder.

be subjected to stress or vibration, a lock joint made by folding back the edges of the stock and interlocking the 2 U sections.

With lap joints, spread flux over the metal areas to be mated and coat them with solder. Then lap the parts and run either a soldering tip or direct flame of a burner over the top of the joint to fuse the coatings. To solder lock joints, clean the contacting surfaces before folding and apply flux. Then, with the parts interlocked, flow solder in, just as you would to sweat a pipe joint.

Handling Rod and Wire
When bonding the ends of rods and heavy wire use lap joint made by filing halfway through each end section. Apply solder to the resulting flat areas, place these "tinned" surfaces together and sweat them with the flame.

Where rods cross, notch both at the point of intersection and sweat-solder the resulting lap joint. Or, if considerable strength is needed, give each rod or wire a half turn around the other and flow solder into the connection.

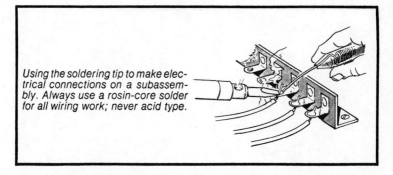

Using the soldering tip to make electrical connections on a subassembly. Always use a rosin-core solder for all wiring work; never acid type.

Electrical Connections

Remove all traces of insulation and oxidation for a good electrical and physical connection. To splice wires, wind the ends together and bend the tips back. Then, holding rosin-core wire solder against one side of the joint, apply heat to the other until the solder melts and binds the whole connection.

Terminals for soldered connections usually have a hole, or "eye." Bend a hook in the end of the wire, place it through the eye, and solder. A pencil flame is good for large electrical connections. For jobs in crowded quarters a small soldering tip is the right tool. Work surely and fast. Prolonged heat will break down the flux, weaken the wire and possibly damage nearby electrical components.

How about brazing? Well, brazing is a first cousin to welding. The chief difference is that brazing is done at lower temperatures than welding and a filler rod is used to fuse the metals together (a filler rod is also used in some welding operations, but its chief function is to fill gaps in the joint).

The rod used in brazing is always of a nonferrous metal with a melting point above 800°F, but always below the melting point of the parts to be joined. As the rod melts, it flows by capillary action between the parts to be brazed. A typical example is the brazing of copper pipe into a tee or elbow. The rod becomes a filler under the action of the heat and flows between the pipe and the elbow making a strong, waterproof connection.

Braze welding—also called bronze welding—also uses a rod as a filler, but the work is of a more gross nature—joining parts by means of fillets, grooves and butt joints. Capillary action is not a factor in this type of work.

The BernzOmatic Super Torch lends itself admirably to brazing operations. It is always used with Mapp (a registered trademark of Arco, Inc.) gas as this gas delivers a hotter flame, 3,700°F, more than 500° hotter than propane gas can deliver. While Mapp gas is somewhat more expensive than propane gas, in the long run it is actually more economical to use. First of all, inasmuch as it burns at a higher temperature, it will complete

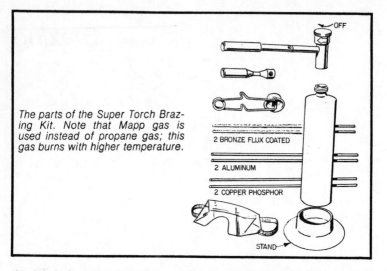

The parts of the Super Torch Brazing Kit. Note that Mapp gas is used instead of propane gas; this gas burns with higher temperature.

2 BRONZE FLUX COATED

2 ALUMINUM

2 COPPER PHOSPHOR

STAND

the job in less time, and secondly, the Mapp gas cylinder holds about 13% more gas than the propane cylinder.

Can the Super Torch be used with regular propane gas? Yes. However, since this torch is not designed for use with propane, there is no advantage when propane is used.

How about using Mapp gas in a regular propane-burning torch? Best not to do so. Conventional propane gas torches are not designed to accept the higher temperatures delivered by Mapp gas. And, incidentally, Mapp gas should never be used with any appliance containing more than 67% copper.

How about cutting or welding steel? Can the Super Torch do these jobs? The answer is no. This torch will braze, weld aluminum (because of its low melting point) and do hard soldering, but welding and cutting are best done with the BernzOmatic Oxygen Torch.

The Super Torch can also be used for flame-hardening, metalizing and annealing. And you can even do ordinary so-called "soft" soldering with the Super Torch. It is indeed a versatile tool.

Household Brazing

There are many, many products around the home that are rendered useless by a simple break in the metal. Garden tools, lawn mowers, hoes, rakes, children's play equipment, wagons,

How to light the torch. Use the sparker that comes with the kit. Crack the valve slightly; too much pressure will prevent ignition.

MAPP

bicycles, swing sets, motor bikes, fences, wrought iron furniture, aluminum pool equipment, barbecue grills and many more—all can be given new life with brazing repair at a fraction of the cost of replacement. In addition, jewelry making, metal sculpture and a host of other hobby and craft projects can be done at home with the Super Torch.

The Super Torch
The Super Torch is different from other BernzOmatic torches. Before you even light the torch you can see that it looks different. The burner unit is positioned at a right angle to the cylinder to permit more comfortable operation. As soon as you light the torch you will *hear* the difference. The cyclone jet action of the flame, which produces the higher heat concentration, actually sounds hotter. And when you begin to use the Super Torch you will quickly see the faster heating speed of the Super Torch.

This torch has been designed to give the maximum usable heat possible from Mapp gas. When used with Mapp gas, the torch will do everything that a propane torch will do, plus brazing —a weld-like bond that is far stronger than common soldering.

To Braze or to Solder
The Super Torch with Mapp gas can be used for nearly any metal-joining job. The choice of brazing or soldering and the

TYPE OF SOLDER OR BRAZING ROD	Stainless steel & flux	General purpose acid core solder	All-purpose resin core electrical solder
APPLY SOLDER OR BRAZING ROD WHEN:	Solder flows freely on contact with heated metal	Solder flows freely on contact with heated metal	Solder flows freely on contact with heated metal
For metals below, use solders or alloys marked "X" at right **ALUMINUM** For strength in joining sheets, sections, etc			
CHROME PLATE For trim, when on steel, brass, copper or nickel alloys (Not on die castings)	X		
COPPER For electrical equipment			X
COPPER OR BRONZE For fittings, tubing, utensils, etc		X	
GALVANIZED IRON OR STEEL For cans, buckets, tanks, eavestroughs, etc.		X	
SILVER AND SILVER PLATE For jewelry, flatware, etc	X		
STAINLESS STEEL For appliances, kitchen equipment or wherever strength is needed	X		
STEEL For utensils, pipes, sheets, tool sheets, motors, etc	X		

UNLIKE METALS such as steel to brass
Unlike metals with X's in the same vertical column can be joined For example copper and galvanized iron with general purpose solder

solder, alloy or brazing rod to be used will depend upon the joint strength required and the metals to be joined. Use the chart on the facing page to select the correct material.

The selection of brazing or soldering will depend on the strength required. As a general rule, soft soldering requires less heat and provides a less strong joint. This process is generally preferred in plumbing and electrical work where the importance of sealing and a good electrical bond outweigh the importance of strength. Brazing, on the other hand, is preferred where the strength of the joint is most important. Brazing produces a joint similar in strength to welding.

SOLDERING AND BRAZING RODS

Aluminum brazing alloy & flux	Silver solder & flux	Aluminum bare brazing rod	Flux coated nickel-silver brazing rod	Flux coated bronze brazing rod	Copper phosphorous brazing rod
Flux becomes a clear liquid	Flux becomes a thin, clear liquid & forms dull red	Brazing rod puddles on contact with heated metal	Brazing rod flows freely on contact with heated metal	Brazing rod flows freely on contact with heated metal.	Brazing rod flows freely on contact with heated metal.
X		X			
			X	X	
			X	X	X
	X	X	X	X	
	X				
	X		X	X	
	X		X	X	

Some Soldering Facts

As indicated in Chapter 1, all solders fall into 1 of 2 categories —hard or soft solders. Soft solders adhere to common metals such as copper, brass, tin and steel. Hard solders must be melted at much higher temperatures than soft solders and thus present a problem at times as their melting points may approach those of the metals to be joined—especially if working on thin stock—but hard solders produce much stronger joints than soft solders.

What makes solder stick? It is a chemical action. At a relatively low temperature, the solder dissolves the surface metal on the parts being bonded and blends with it. This blending can't take place unless both surfaces are completely free from corrosion, rust, grease, paint or any other coating.

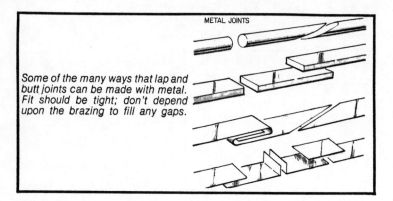

METAL JOINTS

Some of the many ways that lap and butt joints can be made with metal. Fit should be tight; don't depend upon the brazing to fill any gaps.

Why You Must Use A Flux

The moment heat is applied to the work area a new oxide begins to form. This would prevent the chemical action of solder, unless there were a way to float it off. That's what the flux does. Use acid-type flux for heavy soft soldering jobs and rosin-type flux for all electrical connections. Manufacturers of hard solders make special fluxes for their products.

Types of Common Joints

Illustrated are the common types of metal joints. Generally, butt joints are the most difficult (they are impossible with soft solder) and lap joints the strongest. Regardless of the joint to be accomplished or the process to use, proper preparation of the metals to be joined is the single most important part of the job.

Preparing the Work

The importance of proper preparation of the metals to be joined cannot be overemphasized. First, the metal at the area of the joint and any surrounding metal that will come in contact with the filler or brazing metal must be clean, free of grease, oil, oxides and other contaminants. Oxides can be removed by using an emery cloth or sanding or filing.

Next, the metals to be joined should be properly positioned and fastened or clamped firmly in place to prevent movement during the soldering or brazing operation. Be sure that the metals are positioned so that you are able to heat all sides without moving the joint or burning materials in the areas around the metals.

The Brazing Operation

The chart on page 22 will guide you in the choice of brazing rods to use, depending on the metals to be joined and the joint strength required.

As a general rule, bronze brazing is the most versatile and provides the strongest joint. When done properly, bronze brazing is as easy as soldering and produces a bond as strong as welding.

Follow carefully the metal preparation steps. Note that bronze brazing rods often are flux coated. No further fluxing is required.

Heat the metals to be joined until they are cherry red. Try to keep the flame in one place. Moving the flame will allow the metals not in direct contact with the heat source to cool. The tip of the torch should be held about 1/2'' from the metal.

When the base metal is thoroughly hot all the way through the joint (not just at the surface) introduce the brazing rod into the torch flame, touching the joint at the hottest point. Rub some flux from the end of the rod onto the joint. When both joint and rod are hot enough the rod will melt and flow easily and quickly into the joint. Now move the torch along the joint, repeating the same sequence as required.

When using either the copper-phosphorus or aluminum rods, the procedures are similar. However, the metals will be heated to lower temperatures. Copper-phosphorus rods are used frequently in plumbing work. They allow pipes to be joined even though some water may be in them. When using copper-phosphorus brazing rods, heat the metals until they are a dull red color, then follow the procedures described for bronze brazing.

Aluminum brazing is a bit more difficult since the melting point of the brazing rod is quite close to the melting point of aluminum. For this reason it is best to practice a few times on scraps of aluminum before trying a repair job. Use aluminum flux and clean parts to get a good joint.

When brazing aluminum, heat the joint for about 6 seconds. Apply the brazing rod. As soon as the rod begins to flow, remove the heat. Repeat this procedure as needed to complete the repair.

Heat Control

Heat control is often necessary to protect the area surrounding the heated metals or to preserve the temperature of the metals to be joined.

For example, when brazing pipes that are positioned near wood or other combustible materials, protect these combustibles with a heat shield of asbestos or a similar non-combustible and heat-resistant material.

When a large area is to be brazed, the joint should be surrounded with heat blocks or a heat shield to retard heat loss in the metal and reduce the time needed to heat the joint to the desired temperature.

Chapter 3

*Welding–It's Easier
Than You Think*

There are 4 commonly used methods of joining metals together by means of heat. The most familiar, of course, is soft soldering. Soft soldering is used for connecting wires together where a good electrical connection is required and strength is not a factor. It is also used for joining copper pipes and tubing in plumbing work.

Hard soldering is used when a much stronger joint than a soft-soldered joint is required. Hard soldering is used extensively in jewelry making and in the plumbing industry when a particularly strong joint is required. Special solders are used for hard soldering in which silver is the chief ingredient

Brazing is a form of hard soldering; it resembles welding in some respects. The chief difference is that in brazing the 2 metals are joined by another metal (the brazing rod). The brazing rod has a lower melting temperature than the 2 metals to be joined and for this reason the brazing process is somewhat easier to complete than welding.

Welding is a metal-joining process that uses the heat generated by a combination of oxygen and a fuel gas to actually melt the metals so that they flow together and are integrally joined when cooled. Theoretically, a good weld can be made by melting the 2 metals together, but in actual practice a filler or a welding rod is used to fill the gaps and to smooth the finished joint.

Welding requires a flame with an exceedingly high temperature and it is for this reason that oxygen is always part of the gas welding process.

It is interesting to note that welding first became popular a few years after the end of World War I. Prior to that time it was more or less a laboratory novelty and while used in some parts of industry, its acceptance was rather fragmentary, being used chiefly for repair work.

The Treaty of Versailles ending World War I prohibited the German navy from building any vessels whose tonnage exceeded 10,000 tons. In an effort to get around that restriction, the Germans built a so-called "pocket" battleship with *welded* joints. The welding process did away with the thousands of rivets normally used to build a ship and thus enabled the Germans to build the ship within the weight limitations of the treaty. In time, shipbuilders jumped aboard the welding bandwagon and soon most ships were constructed by the new-fangled welding process.

The BernzOmatic Oxygen Torch

Simple welding can be done quite easily with the BernzOmatic Oxygen Torch. It uses the same principles as the commercial oxyacetylene welding torch, but instead of using dangerous acetylene gas, propane or Mapp gas is used in combination with oxygen. (In case you are wondering what Mapp gas is, it is a specially-designed gas composed of liquified stabilized methylacetylene propadiene compound. It is quite a mouthful and you'll agree it is much easier to say "Mapp.")

The BernzOmatic Oxygen Torch will operate just as easily as the conventional propane torch but with its much higher temperature can be used for cutting, brazing, welding and many other jobs around the house that require a high temperature.

Please adhere to the following instructions when using this torch. Take the time to practice on scrap metal to familiarize yourself with its operation. While many safety features have been built into the torch, the exceedingly high temperatures produced can possibly damage the metal you are working with or can even cause personal injury unless instructions are carefully followed and the torch is correctly handled. This torch is a most significant breakthrough for home welding and light industrial use and will undoubtedly be a most welcome cost- and labor-saving addition in your workshop.

Why Oxygen is Important

Oxygen itself does not burn, but when any combustion occurs in the presence of oxygen, the temperature of the combustion process is greatly increased. It is impossible to obtain sufficient oxygen from the surrounding air to produce the temperature required for flame cutting or welding and it is for this reason that "bottled" oxygen is used to supply the extra oxygen needed. The torch is designed so that you can regulate the amount of oxygen to be mixed with the fuel and so control the temperature of the flame—the more oxygen supplied, the hotter the flame. You can therefore control the flame temperature according to the job at hand.

The BernzOmatic Oxygen Torch can use propane or Mapp

gas. As a general rule, Mapp gas will produce a somewhat higher heat output and may allow completion of braze welding or fusion welding jobs sooner and with less oxygen. The choice of fuels is yours. Experience and experimentation will be the best guide. *Never attempt to use any other fuel except Mapp gas or propane.*

Before you can do any welding with the BernzOmatic Oxygen Torch you will need a cylinder of propane or Mapp gas; either one will do but the Mapp gas will yield a hotter flame when combined with oxygen than propane gas. Despite the fact that Mapp gas is somewhat more expensive than propane gas, it really may be more economical to use in the long run as the hotter oxygen-Mapp gas combination heats up faster thus cutting the working time—and less gas will be used.

Before starting your work, there are a few precautions to observe. Avoid inhaling the fumes—work in a well-ventilated area. If indoors, make sure windows are open. Use the torch in an upright position. Extreme tilting may result in an erratic flame and even cause the flame to go out. If the flame does go out, turn off the oxygen and fuel valves.

If the torch starts to hiss with a shrill sound (this means a flashback) turn off both valves and allow the torch to cool before relighting it. Keep grease and oil away from the oxygen regulator valve. Oxygen may cause grease and oil to flame violently.

Never store oxygen and fuel tanks near heat or an open

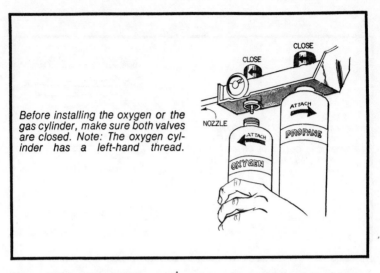

Before installing the oxygen or the gas cylinder, make sure both valves are closed. Note: The oxygen cylinder has a left-hand thread.

flame. When the cylinders are empty, do not incinerate them—they may explode. Read the instructions that come with the propane gas torch.

And, of course, keep the cylinders out of reach of children and never try to inhale the oxygen; it is not made for medical use —it is for welding.

Checking Oxygen Content
When starting any involved welding job, it is best to start with a full cylinder of oxygen. This is easy to determine. If the gauge on the torch assembly reads 400 psi or more the cylinder is full.

Assembling the Torch
Complete instructions on how to assemble the BernzOmatic Oxygen Torch come with the torch. But, in case you have lost the instructions, this refresher should be followed. First, make sure that both valves are in the off position. The oxygen cylinder is screwed into the opening *nearest* the torch tip—it has a left-hand thread (counterclockwise). The Mapp gas or propane cylinder is screwed into the opening nearest the handle—it has a right-hand (clockwise) thread. Mapp gas or propane gas can be used in the opening marked "propane."

Do not use a wrench to install either cylinder; hand tightening is enough. And don't try to be a Samson—tightening with 1 hand is sufficient.

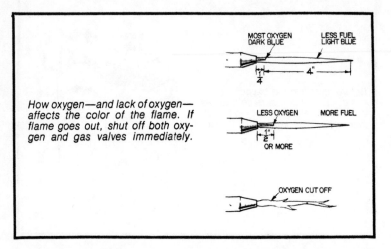

How oxygen—and lack of oxygen—affects the color of the flame. If flame goes out, shut off both oxygen and gas valves immediately.

Okay, now both cylinders have been installed—fuel near the handle, oxygen near the tip. Now, crack the valve serving the fuel cylinder about 1/16th of a turn and light the flowing gas with a spark lighter. Hold the lighter at a slight angle to the side of the torch. It is best not to use a match as the pressure of the gas is apt to blow the match out and the match may burn your fingers before it has had a chance to ignite the fuel.

Now that the fuel is burning, turn the valve to get a 4'' flame. This flame should be in contact with the nozzle of the torch. Opening the valve too much will cause the flame to burn a short distance from the nozzle. Keep the flame in contact with the nozzle. Now gently crack the oxygen valve. Turn this valve to the *left*, in a counterclockwise direction. Turn it to get a sharp, light blue flame.

Adjusting the Flame

As you open the fuel and oxygen valves, variations in the flame pattern will be apparent. For most welding jobs the most efficient flame is a flame with a sharp light blue outer flame with a pinpoint darker blue inner flame. The blue flame should be about 1/4'' in length. The hottest part of the flame is at the extreme tip of the outside flame.

To get a flame of lower temperature, decrease the flow of oxygen and increase the flow of the fuel gas. The inner flame will be longer, 1/2'' or more in length, and will also get wider. This lower-temperature flame is useful for heat treating, annealing, case hardening and heating thin metals.

If you get a long, yellow, pulsating flame, that means that the oxygen supply has been cut off or there is no more oxygen in the cylinder. No flame at all means that the fuel cylinder is empty—or its valve is closed. A flame that goes out or jumps off the tip means too much fuel pressure; close the fuel valve slightly to correct this problem.

Practice Makes Perfect

And this also applies to welding. Practice on scrap metal of the same composition and thickness of the welding you are going to undertake. Remember, too high a temperature could damage the work. Wide-open valves on the oxygen and fuel cylinders will yield the hottest possible flame—and it will also use up both gases at the fastest possible rate. Does the job really need this excessive heat? Probably not. So, turn down both valves to conserve fuel and oxygen. Very few repair jobs require wide-open valves.

Mapp Gas or Propane?

Which to use? It really depends upon the job at hand. If you are working with particularly heavy metals, then use Mapp gas; lighter work can usually be done with propane. Metal cutting, for example, can be done with the less expensive propane gas.

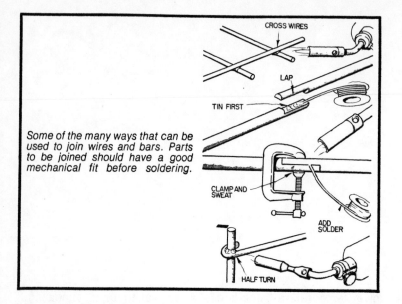

Some of the many ways that can be used to join wires and bars. Parts to be joined should have a good mechanical fit before soldering.

Labels in figure: CROSS WIRES, LAP, TIN FIRST, CLAMP AND SWEAT, ADD SOLDER, HALF TURN

Aluminum less than 1/4'' thick can be "worked" with propane instead of Mapp gas.

Preparing the Work

Before lighting the torch, make sure you have done all of the preliminary work. No sense in wasting fuel and oxygen while you scurry around looking for the right welding rod or clamps to hold the work in place. Goggles, water, and a suitable fire extinguisher should be at hand. Also, make sure the area behind the welding job is protected against the hot flame of the torch. Have pliers on hand to handle hot work. The work should be placed on a firebrick or an asbestos pad. Do not use ordinary brick as trapped moisture inside the brick may cause it to explode when exposed to the high temperature of the propane gas torch.

When everything is in place, thoroughly clean the metal at the joint area. Make absolutely certain it is clean and free of grease, oil, oxidation and dirt. Use an emery cloth, a steel brush or a powered grinding wheel to make sure the weld area is bright and clean.

The next step is to place and clamp the metals to be joined (or the repair job) so that it will be absolutely immovable during the operation. Make certain that the torch will be able to reach all areas of the work that will require heating—and without scorching the adjacent wall or the workbench. The parts to be joined should have a good mechanical fit; if they don't, use filler metal or the welding rod to fill any gaps. The best possible weld is made when the parts to be welded are in good physical contact with each other. Metal will expand as the joint area is heated and it is for this reason that adequate clamping is a must. Clamps, bolts, steel straps and even heavy weights can be used to immobilize the work.

Welding the Work

The welding process really consists of melting the adjacent metals so they form a puddle of molten metal. When the puddle cools, the joint or the repaired area should be as strong as the parent metal. However, in actual practice, a welding rod is introduced into the weld area to fill up irregularities in the weld so that the finished job will present a uniform appearance. See page 22 for the rod to use with different metals.

Light the torch and adjust the flame to get a smooth burning blue tip about 4'' long. Apply the flame to the work (wear goggles and gloves) until the metal in front of the tip starts to puddle or melt. Hold the torch at a 45° angle to the work and point the tip in the direction that the weld is to take. The movement of the torch tip can be one of many patterns: back and forth, a zigzag motion, a circular motion or a combination of these patterns. Choose the method that is most comfortable for you.

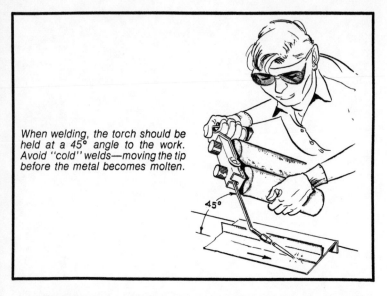

When welding, the torch should be held at a 45° angle to the work. Avoid "cold" welds—moving the tip before the metal becomes molten.

45°

Professional welders use 1 of 2 welding styles: forehand or backhand welding. In forehand welding the welding rod is in front of the torch tip while in backhand welding the welding rod is in back of the torch tip. Backhand welding tends to result in smaller puddles of molten metal and some welders maintain that this is a more efficient way of welding. But if you find that you can do a better and neater job with forehand welding, use the forehand method. The rules are flexible. In both styles, the welding rod as well as the torch is held at a 45° angle to the work.

Welding Without a Filler Rod

A great deal of welding can be done without the use of a filler rod. Position the work so that you can start in at the right (if you are right-handed). Hold the torch tip so that the tip points to the left at about a 45° angle to the work. Now move the flame across the proposed welding area, back and forth in 1/4" movements. As the metal starts to melt, keep moving the torch to the left in a sort of zigzag pattern. Do not move the torch faster than the advance of the molten metal. If you move too fast, the weld will not be strong; if you move too slowly, the torch will burn a hole through the metal. As you near the end of the weld area, lift the torch to avoid burning a hole in the metal.

Check the weld by picking up the work with pliers. Look at the back. Has the weld properly penetrated the rear? It should. If it hasn't (a fault with many beginners), do the weld over. We want you to be perfect.

Galvanized Metal
Before attempting to weld galvanized metal, the zinc coating should be completely removed. Noxious gases develop if the galvanized metal is heated without removing the zinc coating. Filing, sanding or the use of a powered wire brush will remove this coating.

Welding Aluminum
Aluminum melts at a much lower temperature than steel and is rather difficult for the beginner to weld. However, it can be welded—all it takes is a bit of practice. Use scrap aluminum of the same thickness as the work. Successful welding of aluminum is a common, everyday job with the pros—they have the practice, which is all you need. Most trouble with welding aluminum derives from the fact that the operator applies too much heat; about 6 seconds is usually enough, unless working with large castings of aluminum.

Testing the Weld
Want to know how good a job you did? It's easy. Grind the welded area flush to the parent metal (most welds wind up some-

what higher than the surrounding metal). Cut the welded piece in half across the weld. Bend each piece into a U shape; one piece bent with the weld on the inside and the other with the weld on the outside. A good weld will bend without cracking.

Rod Sticking

If the welding rod tends to stick to the work, play the flame directly at the tip of the rod. This will loosen it. Do not try to jerk the rod free. Sticking of the welding rod is due to keeping the rod at the edge of the metal puddle instead of in the middle. The edges of the puddle are near the "freezing" point and tend to grab the rod as it is cooling. While you are freeing the rod, the puddle will undoubtedly be solidifying, so be sure to reheat the metal at that point so a fresh puddle will be formed.

Don't be discouraged if your first attempt at welding looks lumpy. Try again. With a little bit of practice your weld jobs will be the equal of any professional's.

Chapter 4 *Metal-Working Terms*

At one time the author was a copywriter for a New York advertising agency that had a large steel manufacturing company as its client as well as a company that made gases for welding and brazing. As the newest writer in the company I soon realized I had a lot to learn when I was assigned the task of writing copy for both of these clients. Nothing daunted, as they used to say in the old Frank Merriwell novels, I enrolled in a crash course in metallurgy at Brooklyn Polytechnic Institute.

Soon I was familiar with such terms as "austenite," "annealing," "cementite" and "martensite," and ultimately became the resident expert on metallurgy at the agency. I was able to talk to the clients on their terms using their language.

So, whether you are planning to become a professional welder or just want to do welding for your own "amazement," it pays to know what the following terms mean in the welding and metallurgy field.

Anneal deoxidizing: A sub-critical anneal performed in an inert atmosphere in order to minimize oxidation, remove internal strains after cold reduction, decrease hardness and tensile strength and develop maximum ductility.

Annealing: Heating followed by cooling to soften metal and to reduce internal stress.

Austenite: A form of steel produced when carbon and iron are heated above 723°C. Named after a Briton, William C. Roberts-Austen.

Brazing: Joining metals by fusion of nonferrous alloys that have melting points above 800°F but lower than those of the metals being joined. This may be done with a torch (torch brazing), in a furnace (furnace brazing) or by dipping in a molten flux bath (dip or flux brazing). The filler metal is ordinarily in rod form in torch brazing, whereas in furnace and dip brazing the work is first assembled and the filler metal may then be applied as wire, washers, clips, bands or may be integrally bonded, as in brazing.

British Thermal Unit (BTU or Btu): A unit of heat. One BTU is the amount of heat required to raise the temperature of 1 pound of water 1°F.

Butt welding: Joining 2 edges or ends by placing 1 against the other and welding them.

Case hardening: A process of hardening a ferrous alloy so that the surface layer or case is made substantially harder than the interior or core. Typical case-hardening processes are carburizing and quenching, cyaniding, carbonitriding, nitriding, induction hardening and flame hardening.

Cast iron: An iron alloy that contains large amounts of carbon, making it very brittle and unsuitable for forging or rolling.

Cementite: A hard, brittle compound of iron and carbon occurring as a cast iron and steel.

Cyaniding: A process of case-hardening a ferrous alloy by heating in molten cyanide, thus causing the alloy to absorb carbon and nitrogen simultaneously. Cyaniding is usually followed by quenching to produce a hard case.

Decarburization: The removal of carbon from steel by heating it in a medium that reacts with the carbon.

Flame annealing: A process of softening a metal by the application of heat from a high-temperature flame.

Flame cutting: Cutting steel by burning a portion out by means of an oxygen-fuel torch, or removing a part of the surface by means of the torch, more properly called "scarfing."

Martensite: A hard and brittle steel formed by rapid quenching; named after a German, Adolph Martens

Oxide: Usually refers, in the steel industry, to oxide of iron; many mixtures of these oxides form on the surface of steel at different temperatures and give the steel different colors, such as yellow, brown, purple, blue and red. Oxides must be thoroughly removed from the surface of steel objects which are to be coated with tin, zinc or other metals.

Quench hardening: A process of hardening a ferrous alloy of suitable composition by heating within or above the transformation range and cooling at a rate sufficient to increase the hardness substantially. The process usually involves the formation of martensite.

Quenching: A process of rapid cooling from a high temperature by contact with liquids, gases or solids.

Quenching crack: A fracture resulting from thermal stresses induced during rapid cooling or quenching. Frequently encountered in alloys that have been overheated and liquated and are thus "hot short."

Soldering: Joining metals by fusion of alloys that have relatively low melting points—most commonly, lead-base or tin-base alloys, which are the soft solders. Hard solders are alloys that have silver, copper or nickel bases and use of these alloys with melting points higher than 800°F is generally termed "brazing."

Stress relieving: A process of reducing residual stresses in a metal object by heating the object to a suitable temperature and holding for a sufficient time. This treatment may be applied to relieve stresses induced by casting, quenching, normalizing, machining, cold working or welding.

Welding: A process used to join metals by the application of heat. Fusion welding, which includes gas, arc and resistance welding, requires that the parent metals be melted. This distinguishes fusion welding from brazing. In pressure welding, joining is accomplished by the use of heat and pressure without melting. The parts that are being welded are pressed together and heated simultaneously so that recrystallization occurs across the interface. The molten puddle must be protected from any drafts and filler metal must be applied by melting the end of the rod into the molten puddle.

Chapter 5 *Plumbing Work*

Both rigid copper pipe and copper tubing can be used for running hot and cold water pipe lines through the house. Quite often copper tubing is used in combination with rigid copper pipe.

Copper pipe comes in 10' and 20' lengths and in 3 grades: Type M, thin wall; Type L, medium wall; and Type K, thick wall. The thin wall is generally used for residential work. Pipes to be buried or subject to mechanical abuse should always be thick wall.

The diameter of copper pipe is 1/8'' greater than its nominal size. It is available in diameters from 3/8'' to 3''. Working with copper pipe is a great deal easier than working with threaded pipe. For example, if a new length of pipe has to be installed between an existing pipe run, no union is required as with threaded pipe. No threading of *any* kind is required as all joints are made up with soldering.

As with threaded steel or brass pipe, allowances must be made for the distance the pipe fits into its coupling or fitting. A good way to do this is to measure the distance from fitting to fitting and then add twice the depth of the fitting from its shoulder to its end.

A good technique to follow is to make a "dry" run. Lay out the pipe and its fittings, make any necessary adjustments, disassemble and proceed with permanent connections. On long pipe runs, make up and connect 4 or 5 connections at a time and then proceed to the next 4 or 5 connections.

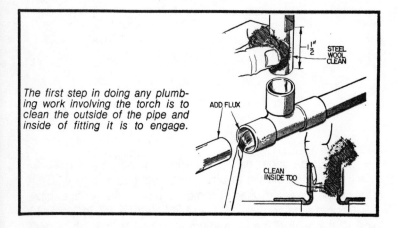

The first step in doing any plumbing work involving the torch is to clean the outside of the pipe and inside of fitting it is to engage.

STEEL WOOL CLEAN

ADD FLUX

CLEAN INSIDE TOO

Making the Connections

The first step, of course, is to cut the copper pipe to the required size. Use a hacksaw with a fine-tooth blade. The use of a miter box will assure you of a square cut. A tubing cutter can also be used to cut the pipe. The next step is to use a half-round file to remove the burr on the inside of the pipe. The hacksaw will make a sort of jagged burr while the tubing cutter will form a fairly even burr. Regardless, make sure all vestiges of the burr are removed. A burr that is not removed will impede the flow of water and, in addition, will sometimes cause a whistling noise in the pipes when a faucet is turned on. So—remove the burr.

The next step is to clean the outside end of the cut pipe with steel wool or fine sandpaper. Clean about 1½'' of pipe. Do the same with the required fitting, but this time clean the *inside* of the fitting. Apply a light coat of soldering flux to the inside of the cleaned coupling or fitting and the same business to the outside of the cleaned pipe.

Push the fitting over the pipe (or the pipe into the fitting, depending upon the layout of the work) until the pipe makes a positive contact with the shoulder of the fitting. Make sure the pipe

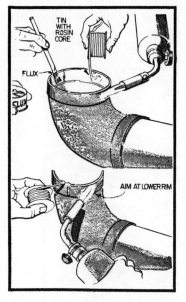

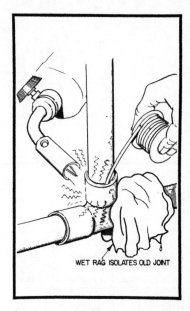

Extra large fittings should first be tinned to make sure a good bond will be effected during the subsequent soldering operation.

Next, apply the torch to the joint. Start feeding the solder only after the joint is hot enough to melt the solder. Note the use of the wet rag.

and its fitting are secure and will not be subject to movement during the next operation—soldering the connection.

Light the propane gas torch and heat the tubing about 1/2'' from the end of the fitting. Keep playing the flame at right angles to the tubing—and keep the flame in motion, gradually heating the fitting as well. After the pipe and its fitting have been thoroughly heated with the torch, apply the solder. If the metal is hot enough, the solder will melt freely and will be drawn into and around the joint by capillary action. The solder will flow into the fitting even if the pipe and its fitting are in a vertical position. Flow solder all around the joint so that the solder will make a fillet around the fitting.

Any excess flux or discoloration can be removed by wiping the joint with a rag while the connection is still warm.

When more than 1 connection is to be made at a point, such as when making a tee connection, all the joints should be made at the same time. However, if this is impossible, make up the first joint and wrap it with a damp cloth so the nearby heat will not affect it.

An improperly made-up joint (Heaven forbid!) can be dis-

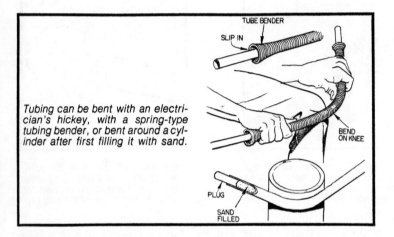

Tubing can be bent with an electrician's hickey, with a spring-type tubing bender, or bent around a cylinder after first filling it with sand.

TUBE BENDER

SLIP IN

BEND ON KNEE

PLUG

SAND FILLED

mantled by heating the joint and gently tapping it or by pulling at the pipe while the solder is being melted.

Never try to make up a connection with water in the pipe. The water in the pipe will dissipate the heat, making it impossible to heat the connection hot enough to melt the solder. But if there is a little water in the pipe due to a faulty shut-off valve that allows slight leakage, do what many pros do—stuff the pipe with some white bread to absorb the water. When the connection is finished and the water is turned on, the bread will be flushed out through the faucet without harm, and the connection has been made despite a dribble of water in the pipe.

Copper Tubing

An important advantage when using copper tubing is its flexibility and its length—it comes in 15', 30' and 50' rolls. Thus, it is often possible to do an entire run between source and faucet with a single fitting except for the fittings at each end.

Copper tubing is made in two types: Type L for most residential use and Type K for hard service such as in commercial establishments. Copper tubing is ideal for remodeling work as it can be snaked through small openings and behind walls without extensive demolition work. However, it does not look quite as neat as copper pipe with its straight runs and exact right-angle turns. But if it is behind a wall, who's looking? Copper tubing can be joined the same way as copper pipe—by soldering or with the use of flare-type fittings.

Inasmuch as quite a bit of bending is part of the work when using copper tubing, good bending technique is important in order to avoid kinking the tubing. One way of bending copper tubing is by means of a spring-type tubing bender. The tubing is inserted within the coils of the spring—they come in various sizes—and the tubing is bent over a curved surface. Your knee works fine for this job. When making the bend, bend it just a bit more than required and then ease it back to the required radius. To remove the tube bender, twist it slightly as you pull it off. To make a really sharp bend in copper tubing without kinking it, fill the tubing with sand before bending it. Make the bend and force the sand out with a garden hose.

Most gutters and downspouts in present-day homes are made of metal, copper, steel or aluminum with a small percentage made of plastic. Some homes may even have gutters of wood—usually as part of the overall design of the house specified by the architect. But, by and large, present-day homes have metal gutters—and they require maintenance and repair.

If your gutters and downspouts are of galvanized steel—the only type made of steel—do not paint them until they have weathered for at least a year. Paint is sure to flake off galvanized metal unless it is first coated with a special primer or allowed to weather for a year or so.

If your gutters are of copper, then of course they will never require painting. They are virtually corrosion-resistant and in time will weather to a dark brown or may even develop a light green patina.

The first indication that average homeowners have that gutters need attention is a cascade of water flowing over the gutters. For some perverse reason, this unwanted water always appears over the front or rear entrance of the house—never where it will not inconvenience anyone. But that is the nature of inanimate mechanical devices—they seem to have a perverse nature in frustrating man.

What's the trouble? Simple. The gutter has developed a sag, either due to a heavy accumulation of soggy leaves or because the hangers which hold the gutter to the roof have loosened or are lost.

If you look at your neighbor's house, whose gutters are not weeping, you will notice that they are not level with the edge of the roof. They have—or should have—a definite slant towards the downspout. This slant is about 1/8'' for each foot of gutter length. If the house has a downspout at each end, then the gutter will slope down to *each* downspout with its high point at the middle of the house.

Now then, to get back to your home. Nothing much you can do now with the rain coming down in buckets. You have a legitimate reason to procrastinate. The next day, or the next weekend, clean the debris out of the malfunctioning gutter—and in all the gutters around the house. Examine the gutter hangers.

Gutters are supported by 1 of 2 methods. One style has a sort of strap which is riveted, or soldered or clipped to the gutters with the other end of the strap nailed to the roof and covered by the roof shingles. The more common—and modern—method is by means of a spike that passes through the top lip of the gutter, the rear of the gutter and into the wood trim immediately below the roof. A sleeve within the gutter prevents the gutter from being deformed as the spike is driven home. Needless to say, these sleeves and spikes should be of copper if your gutters are of copper. Any other metal—aluminum or steel—will create an electrolytic action between the gutter and the spikes causing rapid deterioration and corrosion.

If the strap has pulled away from its rivet, your best bet is

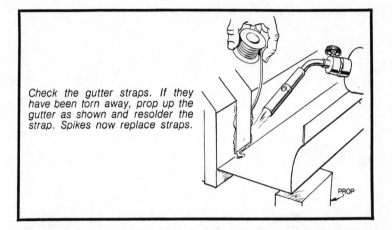

Check the gutter straps. If they have been torn away, prop up the gutter as shown and resolder the strap. Spikes now replace straps.

PROP

to solder the strap back to the gutter. Clean away all debris—which you should have done by this time—and thoroughly sand the bottom of the strap and that part of the gutter where the joining is to take place. Sand until the metal is shiny and bright. This goes for the strap as well as the gutter. Set the strap around the gutter. If there is any chance that the strap will move during the soldering operation, use a small C-clamp to keep it in place. *It is absolutely vital that the strap not move during the soldering operation.*

Fluxes

Before starting any soldering operation, you should know something about fluxes. Fluxes are special chemicals which protect and prevent the freshly cleaned metal from oxidizing. It is virtually impossible to solder any metal that is even slightly corroded. Corrosion is a form of advanced oxidation and oxidation starts within minutes after any metal has been cleaned by sanding or with steel wool. There are many fluxes for use with different metals. However, for our purpose, inasmuch as we are planning to solder heavy copper, muriatic acid can be used as a flux. Use a 1/2'' stiff bristle brush to apply the acid to the area to be soldered, after the metal has been sanded clean. Or you can use a paste flux—but flux you must.

Apply the flame of the propane gas torch to the far side of

the joint, wait until the metal is hot, apply solder, and if the solder melts and starts to flow, proceed to feed more solder while at the same time moving the flame and applying solder until the joint is secure. To remove any excess flux, wipe the joint with a rag before the metal has cooled.

So far, so good. The next step is to check to see if the gutter has a slope toward the downspout. You can check this with a level or by pouring some water into the gutter and noting whether it flows toward the downspout. If it does, well and good; if it doesn't, you will have to adjust the strap so the gutter will have the right pitch.

Carefully lift the shingle under which the strap is nailed and pry out the nail or nails that secure the strap to the roof. Push the strap up to a new position so that the gutter will have the required slope and nail the strap back in place. Apply some roof cement to the nailheads and to the bottom of the shingle that you have pried up.

Lower the shingle to its original position. If the shingle refuses to stay put, place a weight on it. However, if you have done this operation during warm weather, you should have no problem.

The Loosened Joint

Another common problem with gutters is a steady drip from a particular part of a gutter, long after the rain has stopped. During a rainstorm, the drip isn't noticeable as the steady rain hides

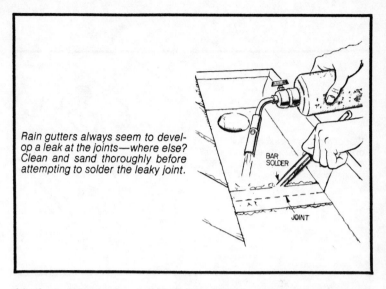

Rain gutters always seem to develop a leak at the joints—where else? Clean and sand thoroughly before attempting to solder the leaky joint.

this fact. What's the trouble? The joint between 2 sections of guttering has loosened, usually because the original installer did not do a conscientious soldering job. It is up to you, now, the harassed homeowner, to do a better job. It is a common axiom that when a person is doing a job for himself, he will do a much better job than if he is doing it for someone else for pay. And now it is your turn to prove this bit of wisdom.

The next fair day, clean out the gutter. Make sure you have wiped away the slightest trace of moisture. Sand thoroughly, or use steel wool for about an inch or so on either side of the break in the joint. As soon as you clean away this area, apply the flux, acid or paste. Bar solder is best for this job, as it costs a good deal less than the solder that comes in wire form and a good deal of solder is used in this operation.

Light the propane gas torch, apply the flame to the work, wait and apply the solder. If the solder flows freely, the metal is hot enough. If the solder sort of wrinkles, the metal is not hot enough. When doing this work, make certain that the 2 ends of the gutter are absolutely immovable during the soldering operation. Any movement will not only weaken the joint, it will result in no joint at all. A C-clamp can be used to hold the gutters together without moving while you are working at one end. Then remove the C-clamp and proceed to finish the job at the other end.

Remember that solder in itself is not a very strong material. So make certain that after you have finished mending the joint

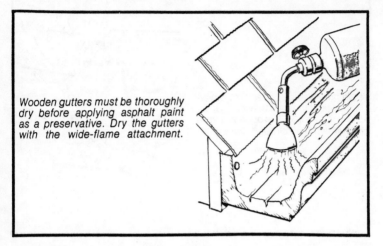
Wooden gutters must be thoroughly dry before applying asphalt paint as a preservative. Dry the gutters with the wide-flame attachment.

the gutter will be securely supported, especially near the joint area. Inspect the hangers. Are they all doing their share of the work? If not, adjust as indicated before.

Aluminum Gutters
Okay so far for copper gutters. But suppose the gutters on your home are made of aluminum. What then? The same business applies, but you must use a special flux and solder especially designed for use with aluminum.

Galvanized Steel Gutters
The technique is similar to working with aluminum. Clean and flux. Use an acid core solder and make sure the metal is thoroughly hot before applying the solder. You can tell when the metal is hot enough by the action of the solder. If the solder flows freely when it is applied to the work you can be reasonably certain that the metal has been sufficiently heated.

Wood Gutters
These gutters require periodic maintenance as sun and rain really wreak havoc with them. Paint the outside of these gutters with a good grade of paint made for exterior use. Best to use alkyd paint rather than the water-based latex paint. How about the inside? The best way to preserve the inside of wood gutters is by an application of roof paint. This is a black asphalt paint

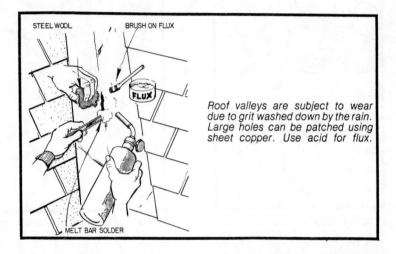

STEEL WOOL　　BRUSH ON FLUX

FLUX

MELT BAR SOLDER

Roof valleys are subject to wear due to grit washed down by the rain. Large holes can be patched using sheet copper. Use acid for flux.

that should be applied only in warm weather and to gutters that are *thoroughly* dry. Best to do this job after several days of hot, dry weather. If you are impatient to start the job, use a propane torch to dry the inside of the gutters. Apply the flame lightly over the inside of the gutter. Remember, you are trying to dry the wood, not burn it.

Apply the asphalt paint with a 3'' brush and give it a second coat after 2 days' drying time.

Valley Flashing

A mysterious roof leak can sometimes be traced to a hole in the flashing. Flashing is usually sheet copper used wherever a roof meets the chimney, a wall or an adjacent roof. In the average home it will always be found in the valley between 2 roofs. Valley flashing leaks are often found near the low part of the valley. As the rain pours down the valley it brings with it a certain amount of gravel and debris dislodged from the shingles. In time, this debris actually wears a hole in the flashing and, before long, rain penetrates the hole, and sure enough you will encounter that familiar wet spot on the ceiling.

Small holes in flashing can be easily repaired by filling the hole with solder. Clean the area around the hole by sanding and apply flux. Use the propane gas torch to fill the hole with solder. If the hole is fairly large, then it is best to make a patch using the same metal as the valley flashing. Clean and sand all around the

Another method of repairing a leak in a valley is by "leading." This consists of applying a layer of solder over the worn-out section.

The crumbling flashing compound around the chimney should be removed by first heating with torch then scraping away with putty knife.

hole area. Clean the patching material, apply flux and solder all around the patch as shown in the drawing.

Leading

Another method of repairing a leak in flashing is by "leading." This repair can be done in copper or galvanized gutters; it is an easy fix-it method. By definition, this means "covering, filling or framing metal with lead or solder." In other words, you don't resolder the joint; you simply flow a waterproof patch of solder over the section that leaks.

Start by wire-brushing or sanding all around the area until the metal is shiny bright. With galvanized steel you must not only remove every trace of paint and rust, but rough up the zinc coating until you've bared the steel beneath.

Next, cover the cleaned surface with acid flux and place the end of a piece of bar solder against the joint, well back from the section that's opened up. Apply the brush flame of the torch to the joint, just ahead of the solder. Wait until the heat of the metals have melted a fair amount of solder. Then move both flame and solder along the seam, slowly enough to leave a continuous, joint-spanning patch. Finally, scrub off the flux residue.

Another place for possible roof leaks is defective flashings around chimneys and vent pipes. Remove the old, cracked flashing compound with a chisel and hammer. You can ease the

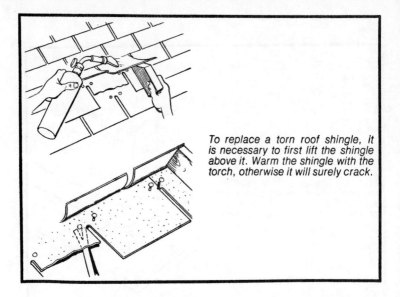

To replace a torn roof shingle, it is necessary to first lift the shingle above it. Warm the shingle with the torch, otherwise it will surely crack.

job considerably with the flame of a propane gas torch. Play the flame over the old flashing compound until it is soft, then scrape it away with a putty knife. Apply new flashing compound, black asphalt roof cement—making sure that all open areas have been completely filled. Again, best to do this job during warm weather.

Missing Shingles

While you are up on the roof, look around for torn, loose or missing shingles. These are a prime source of roof leaks. To replace a missing or torn roof shingle, the first step is to raise the shingle above the damaged one—without cracking it. Best way to do this is to warm the shingle with the torch—warm it, don't burn it. When it is nice and warm, raise it gently and remove the nail or nails under the shingle so you can remove the damaged shingle. Take the nails out with a pry bar, or use a chisel and hammer, but take them out. If you leave them behind, they may work their way up and cause future trouble. Now then, slip the new shingle underneath the raised shingle. Slide it up far enough so that its edge will be in line with the adjacent shingles. If you have any difficulty sliding it into place, trim the corners.

The next step is to nail the new shingle in place. Hold up the covering shingle, or pry it up with a length of wood, and apply your talents and a couple of roofing nails to the new shingle. Use special roofing nails with broad heads; galvanized nails are best.

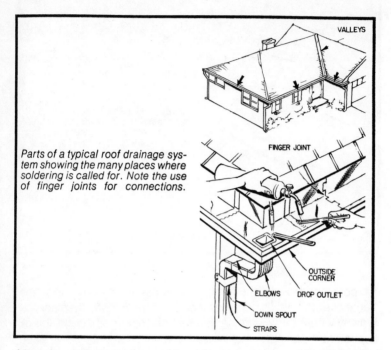

Parts of a typical roof drainage system showing the many places where soldering is called for. Note the use of finger joints for connections.

After the nails have been driven home, apply a dab of roofing cement to the top of each nail head.

Installing New Gutters

The time may come when you decide that the old manse now needs new gutters. This decision usually occurs after a heavy snow or ice storm that has torn down several sections of gutter and inasmuch as the rest of the gutters are in pretty bad shape due to age, falling branches and missing brackets, the big decision is to replace.

Fine (as the gutter manufacturers would say), but what kind? If your old gutters were of galvanized steel, the chances are that you are pretty well fed up with the rusting qualities of steel. Aluminum has its advantages; it is light in weight and costs less than copper. Wood is used as a replacement only if the decor of the house calls for wood. Plastic is a newcomer. It will never corrode and never needs painting. But proper installation is most important with plastic guttering. If plastic gutters are not properly installed, they may buckle in hot weather due to expansion. They also lack the stiffness of steel and copper.

Obviously, the best choice is copper. Measure the area around the house that takes guttering. Bear in mind that most guttering is sold in 10' lengths, so figure out exactly how many lengths you will need. Next, determine how many downspouts the house will require. Just count how many downspouts the old gutters had, or if this is an installation on a new house, figure 1 downspout for each corner of the house, plus 2 extra downspouts where the roofline ends (such as over a window or a door) and continues beyond. You should have 1 downspout for every 35' of gutter. You will probably need some offset elbows so the downspout can be mounted as flush as possible to the walls of the house.

In addition, you will need outside corners, drop outlets, elbows, caps and, of course, nails, spikes and sleeves. Best way to handle this is to make a sketch of the roof perimeter, bring it over to the roofing dealer and listen to his suggestions. Make certain that you will have a support for every 3' of gutter run. Skimping here will cause you grief later—the gutter will sag.

Well, everything has been delivered to your driveway, or else you have picked it up, and now what? Start at one corner of the house. Install the first section—which will have an end cap. This section with an end cap will be the high side of the gutter. In other words, all water which collects in this section will flow down to the next section. You can use a string, a level or a ruler to make certain that this first length will be sloping towards the low end. A slope of 1/8" per foot should be maintained for good drainage.

If the gutters will be supported by spikes—the most common method—drill holes in the front and back of the gutter for the spikes. If you have to cut a length of gutter to fit, support the inside of the gutter with a length of 2x4 and use a hacksaw with a fine-tooth blade for cutting.

Professional roofers will install the complete system of gutters and downspouts before doing any soldering. So, you may as well learn from the pros and do likewise. Install each length, section by section, overlapping each section by an inch or so. As further support, many roofers will make several cuts in each end of each gutter so that the "fingers" of the gutter end will engage the "fingers" of the next length of gutter. Only after the complete installation has been made is the next and most important step to be taken—soldering all joints. Use sandpaper, apply acid or paste flux and the propane gas torch to make a secure joint between each section.

Downspouts

Downspouts should give you no trouble. The downspout slips over the elbow or the drop outlet. *It should not be soldered in place.* A strap around the downspout should be nailed to the wall to keep the downspout from shifting. The bottom end of each downspout leads into a dry well or a concrete splash pan to divert the water away from the house foundation. Incidentally, water in the basement at a point where a downspout is located is often an indication that the dry well at that particular point is no longer doing its job or has collapsed. Instead of digging up your lawn and installing a new dry well, lead the water from the downspout to a point at least 6' from the house wall by means of a second downspout. The author had this annoying problem and solved it by having this particular downspout discharge its water into the driveway, and thence into the street. Worked like a charm—the basement is now dry, whereas formerly after a heavy rain a puddle was sure to collect at the corner of the basement facing the defective dry well.

Chapter 7

Installing Floor Tiles

Covering a floor with tiles is a sure-fire method of sprucing up a room. Floor tiles are made of asbestos, asbestos-vinyl, all vinyl and cork. The asbestos and vinyl-asbestos tiles are the ones to use for below-grade application, though they can be used above grade. But the vinyl and cork should not be used below grade.

Tiles come in 1/8" and 1/16" thicknesses and in 9" and 12" squares. Measure the area to be covered. Length times width will give area in square feet. If you are planning to use 12" tiles, then that is the number of tiles you will need; for 9" tiles, use the table.

Tiles are available in dozens of colors and patterns. For easy maintenance, avoid solid colors. Make sure you have the following tools on hand: a steel tape, string and chalk, pencil, a sharp knife, an adhesive spreader, sandpaper, a plane, a vacuum cleaner and a propane torch.

Remove all furniture and pry off any molding at bottom of the baseboard. Any high spots, bumps or ridges on the floor? Plane them down. Any cracks between boards? Fill them with wood putty, let dry and sand.

If the floor has been painted, sand it. Loose floor boards should be nailed down. Nails should be flush or slightly recessed. Floors in poor condition should be covered with hardboard. Hardboard for underlayment comes in easy-to-handle 4' squares. Nail the panels in a staggered fashion so there will be

To find the number of 9" tiles required for a room, determine the length and width. Find the length and width in the table above, follow them down and across to the point where they intersect. For example, you will need 252 9" tiles to cover a floor that measures 13' x 10'.

Feet	1	2	3	4	5	6	7	8	9	10	11	12	13	14	15
1	4	6	8	12	14	16	20	22	24	28	30	32	36	38	40
2	6	9	12	18	21	24	30	33	36	42	45	48	54	57	60
3	8	12	16	24	28	32	40	44	48	56	60	64	72	76	80
4	12	18	24	36	42	48	60	66	72	84	90	96	108	114	120
5	14	21	28	42	49	56	70	77	84	98	105	112	126	133	140
6	16	24	32	48	56	64	80	88	96	112	120	128	144	152	160
7	20	30	40	60	70	80	100	110	120	140	150	160	180	190	200
8	22	33	44	66	77	88	110	121	132	154	165	176	198	209	220
9	24	36	48	72	84	96	120	132	144	168	180	192	216	228	240
10	28	42	56	84	98	112	140	154	168	196	210	224	252	266	280
11	30	45	60	90	105	120	150	165	180	210	225	240	270	285	300
12	32	48	64	96	112	128	160	176	192	224	240	256	288	304	320
13	36	54	72	108	126	144	180	198	216	252	270	288	324	342	360
14	38	57	76	114	133	152	190	209	228	266	285	304	342	361	380
15	40	60	80	120	140	160	200	220	240	280	300	320	360	380	400
16	44	66	88	132	154	176	220	242	264	308	330	352	396	418	440
17	46	69	92	138	161	184	230	253	276	322	345	368	414	437	460
18	48	72	96	144	168	192	240	264	288	336	360	384	432	456	480

no continuous seam. Use nails with annular rings. After all the filling, planing and sanding have been finished, vacuum the floor.

Then measure the two opposite walls for their midpoints and snap a chalkline between these points. Do the same for the other 2 walls. Where these lines intercept will be the *approximate* starting point for laying the tiles.

If you started laying tiles from this point, you may wind up with a border of a full tile at one side and a half tile at the other. We want an *even* border at opposite sides. To get this even border is a matter of shifting "dry" tiles.

Lay out 2 rows of uncemented tiles from the crossed lines to the wall. The chances are that the last tiles will not fit into place unless the final space at each side is exactly 9". No such luck! What we are trying to get is an even border at opposite walls, not a 1" border at one wall and an 8" border at the other—nor do we want a 9" border.

Shift the center line so that the space between the last tile and the wall will be *about* 4 1/2". *Borders on opposite sides of the room should be equal in width*, though not necessarily the same width as the borders on the other two walls.

Working out from the center of the room, apply the cement to an area that will cover 6 tiles or so. Place the first tile down; don't slide it, if you do some cement is bound to crawl up along the edges of the tile preventing good edge-to-edge contact. Lay the next tile down, making sure that the pattern of this tile is at

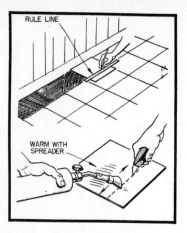

The last tile in a row will always require trimming. Warm the tile with the wide-flame attachment of the torch before cutting the tile.

To install tiles around pipes, cut a slit and a hole equal to the diameter of the pipe. Warm the tile before attempting to cut the hole.

right angles to its neighbor—unless your plan calls for a pattern running the same way.

Now you have come to the wall. And the space calls for a tile that is, let's say, about 4'' wide. But you want this tile to be a snug fit against the wall, no gap please and not too wide.

You will need 2 tiles for this operation. Lay 1 tile over the last tile laid so that you have a double tile at this point. Lay the second tile over the sandwich, making sure that it touches the wall. Draw a line on the top tile using the edge of the second tile as a guide. Inasmuch as asbestos and vinyl-asbestos tiles are hard to cut unless quite warm, use the propane gas torch to warm the tile. Play the flame of the torch over the area to be cut until the tile feels warm to the touch. Now use a straightedge and a knife to cut the tile at the pencil mark. Cement the cut-off piece into place against the wall. The cut-off piece will be a perfect fit against the wall. Continue in this fashion until the entire border area is completed. When you are finished with a quarter of the floor, proceed with the rest of the floor.

Suppose there are some pipes in the room—risers for heat and pipes for hot and cold water. What then? Cut a piece of stiff paper to the exact size of a tile. The folders used for filing letters are ideal for this purpose. Use the paper or file folder as a pattern. A slit directly behind the pipe will allow for a tight fit. With a bit of trial and error you should be able to make a pattern that will fit exactly around the pipe—or any other obstruction, for that matter.

When you are satisfied that the pattern fits as closely as possible, lay the pattern—some folks will now call it a template—over the tile and mark the outline with a soft pencil. Light the propane gas torch, warm the tile along the pencil lines marked as a guide for cutting. Cut away the unwanted part with a sharp knife. Don't hurry this part of the job; after all, you are working for yourself and time is not of the essence.

The use of the propane gas torch is especially important when laying tile on a concrete basement floor. This is so because the concrete in the basement is seldom, if ever, absolutely flat. It is bound to have high and low spots and inasmuch as asbestos and vinyl-asbestos tiles are quite stiff and subject to cracking if bent, heating the tiles with a torch before application will allow the tiles to fit the contours of the floor, thus making a much neater installation than if the torch were not used.

Lifting Old Tiles

Quite often the occasion arises where a damaged tile must be removed (which is why it is a good reason to keep extra tiles on hand that match the original installation). The best way to remove a damaged tile is to heat it gently with the propane gas torch. Use the wide flame attachment on the torch and play the flame over all areas of the tile until it feels uncomfortably warm to the touch. Score a line on all sides of the tile. Next use a putty knife to lift one corner of the tile. Keep moving the putty knife along the edge of the tile until the entire edge has been lifted. At about this time you may find the going a bit tough as the tile has

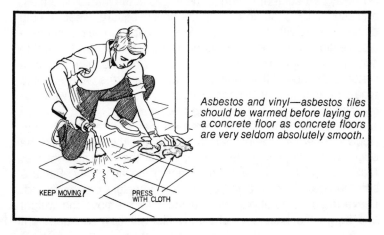

Asbestos and vinyl—asbestos tiles should be warmed before laying on a concrete floor as concrete floors are very seldom absolutely smooth.

KEEP MOVING! PRESS WITH CLOTH

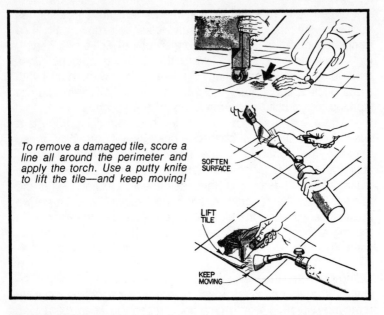

To remove a damaged tile, score a line all around the perimeter and apply the torch. Use a putty knife to lift the tile—and keep moving!

SOFTEN SURFACE

LIFT TILE

KEEP MOVING

now cooled somewhat. Apply more heat and keep shoving the putty knife under the surface of the tile until it is completely lifted.

At this point your job is only half done. Play the torch over the floor surface—be careful not to heat the adjoining tiles—to keep the cement in a softened condition. Again use the putty knife to scrape away the cement. This part of the job may be tedious, but it is most important to remove *all* of the cement before the replacement tile can be installed.

Pay particular attention to the edges of the 4 adjoining tiles, and especially to the corners of the hollow square. Keep the old cement warm with the torch and keep scraping away until every last vestige of cement has been removed. Be careful not to scorch the floor. The final clean-up can be done with warm water and soap—if the original cement was of the water-soluble type—or with a paint thinner. *Be careful not to get the water or paint thinner under the adjoining tiles.*

Next step is to wipe the area so that it is thoroughly dry. Apply the cement, being careful not to get the cement on the edges of the adjoining tiles. Now, install the replacement tile. Make certain that the pattern is facing the right way. Place a smooth board over the tile and a heavy weight on the board. Leave it alone, go about your business and remove the weight and board the next day. The new tile will have an apparent

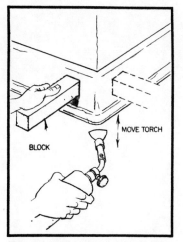

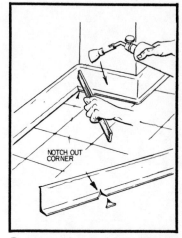

How to make an exterior bend on cove molding. Warm molding and at the same time keep pressing it against the wall as shown here.

To make an interior bend with cove molding, first cut out a notch at bottom of the molding. Then warm the molding and press into place.

freshness compared to the other tiles which have naturally aged. Don't despair; in time the new tile will gracefully age and look like the rest of the tiles in the room—scratched and worn.

Cove Molding

The application of cove molding is quite easy. It comes in 6' lengths and cuts easily with a knife. The only precaution to take with this molding is to try to make the joints where they will be least visible, such as at that part of the wall that will be hidden by furniture.

Apply the cement to the back of the molding—some styles are lightly ribbed to give the cement better adhesion. Install the molding so that the bottom curve is in good contact with the floor tiles and the back, of course, making good contact with the wall.

How about corners? Here is where the **propane gas** torch comes in mighty handy. Use the torch flame to warm the molding. This will enable you to make a much tighter right angle bend when making an outside corner. Use the torch after you have applied the cement and make the installation while the molding is still warm.

To install the molding where an inside corner is involved, the technique is slightly different. Make a dry run and determine where the bend for the inside corner will occur. Then cut out a small V at the bottom of the molding where the bend will be. Heat the molding with the torch and apply to the wall surface. If you have made the V cut properly, then you should have a perfect miter joint at the bottom of the molding where it touches and meets the tiled surface. If there is a slight overlap, then the V has not been cut wide enough. Do not despair. Widen the cut with a sharp knife or a single-edge razorblade to make the correction and all is well.

Go over the entire molding with a clean rag, pressing the rag against the molding to make certain that there is good adhesion between the molding and the wall.

Tools such as chisels and axes may lose their hardness because of frequent resharpening and consequent overheating. Steel, especially carbon tool steel, can be hardened by heat treatment. The process consists of heating the metal to a cherry red color. Then cool it by quickly dipping it into oil or water. To make sure of even hardening, the oil or water should be thoroughly stirred while the metal is immersed in it.

Cold-rolled steel, machine steels and any other steel having a carbon content of less than 3% should not use the above process. Instead they should be *case-hardened*. To case-harden steel, use case-hardening powder sold under a proprietary name. First heat the steel to a cherry red and roll or immerse the steel in the compound. Then heat the steel again to a cherry red and quickly quench it in water. This type of operation forms a hard case on the surface of the metal and this is why it is called case-hardening.

Tempering Steel

This is a sort of reverse process; instead of hardening the steel, it may be necessary at times to reduce its extreme hardness and brittleness. Use the BernzOmatic Oxygen Torch, or Super Torch to heat the object in accordance with the accompanying chart.

Color	Temperature	Tools
Faint yellow	420° F	knives, hammers
Light yellow	440° F	lathe tools, scrapers
Straw	460° F	bits, reamers, punches, dies
Light brown	480° F	large taps, twist drills
Dark brown	500° F	drift pins, chisels, axes
Purple	540° F	center punches, cold chisels
Blue	560° F	gears, springs, screwdrivers
Dark blue	600° F	spokeshaves, scrapers

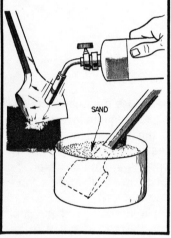

A cold chisel can be annealed by heating the edge to a cherry red color and then by burying the tool in sand so it will cool very slowly.

To restore the temper of an edged tool, heat it as shown, and then plunge it into water. Excessive grinding draws temper out of tools.

Before applying the torch to the steel, clean the surface thoroughly; if you don't, the color will be hard to distinguish. Heat the object slowly and when the correct temperature, indicated by the color, has been reached dip the tool or item quickly into water.

Annealing
Annealing is a process similar to tempering, but it is used to "soften" metal. Copper and brass are annealed by heating and

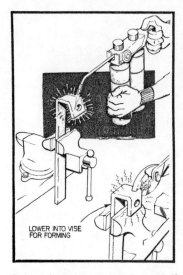

The struck end of a tool should always be annealed so that it is softer than the face of the striking hammer to avoid possible chipping.

It is quite easy to bend metal if it is first heated. Heat and bend in stages until the desired shape is obtained. Heat to a cherry red.

then cooling quickly by dipping in water. Steel, however, should be cooled slowly. Heat the steel to a cherry red color and then cover the steel completely with sand so that it will cool slowly. The longer the cooling process takes, the better the job will be. This cooling operation, depending upon the size of the item, may take from 4 to 24 hours.

Bending Metal

All very well to anneal, temper and harden steel, but how about just bending a particularly heavy length of steel or strap iron? For this operation, in addition to the torch, you will need a hammer and a vise. Mount the steel to be bent in the vise with the proposed bending area a few inches above the vise jaws. Light the torch—the BernzOmatic Oxygen Torch is best for this purpose—and play the flame on all sides of the area to be bent.

When this area is cherry red, loosen the vise slightly and use pliers to lower the work so that the bend area will be as close to the vise jaws as possible. Tighten the vise jaws securely and use a hammer or a light sledge to pound the red-hot metal into the desired bend. Light pounding will usually be sufficient as the heated metal will be quite willing to comply with the dictates of the hammer blows.

A particularly heavy piece of metal may require additional reheating as the operation is in progress. Again, loosen the vise and position the work so that it is somewhat above the vise jaws. The reason for this is that you do not want to affect the finish of the vise jaws, and also because heavy vise jaws will tend to absorb a good deal of the heat that should be absorbed by the work.

Bending conduit

Whether you're making a 90° elbow in a thin-wall conduit for electrical work or using this handy material for a youngster's jungle gym or go-cart chassis, preheating the metal makes controlled bending easy.

First, drill a hole the same diameter as the conduit in a block of wood. Then saw along the centerline, to divide the block into two vise pads. Place these pads in a metal vise, with the conduit between them. Next, slide a length of BX cable through the conduit to prevent kinking. Apply the flame of the torch to the section to be bent. Move it slowly back and forth several minutes, distributing uniform heat. Then slide a piece of heavy pipe over the free end of the conduit, and pull it to form the bend. Quench the heated section in water and withdraw the BX cable.

Chapter 9 *Working With Glass*

Crash! Okay, one of the kids has batted a baseball through the kitchen window. You can call a glazier to do the job—and in my neighborhood the cost is $13 for a 10'' x 14'' pane of glass installed—with the installer due 2 days later. Honestly, that's the experience I once had. I went down to this same glazier and bought a sheet of 10'' x 14'' glass for $1 and that even included some glazier's points and a wad of putty.

First things first. And the first thing to do is to remove the balance of the broken glass from the window. Use gloves and pry out the glass, piece by piece. Sometimes it helps if you break the large pieces with a hammer blow—stand clear and wear goggles.

Next step is to remove all the dried putty in the window frame. This may be a bit difficult as its aging process started several years ago and it is literally as hard as a rock. Light up a propane gas torch with a pencil-type burner and play the flame on the putty, just enough to soften it, not to scorch the window frame. Work on a small area at a time, using the torch as needed until all of the putty has been removed. Use a putty knife and screwdriver to remove the putty. Next, sand the inside of the frame until it is smooth and free of bumps. Now it is safe to measure the opening.

Unless you are adept at glass cutting, and you have some large pieces of scrap glass in the basement or garage, your best bet is to hie yourself down to that independent glazier and buy it

Removing hardened putty to replace a broken window pane is a fairly easy job if you first soften the putty with the flame of the torch.

How a chipped edge can be eliminated from a crystal goblet. Keep flame of the torch on the chipped area until it glows a bright red.

from him. But ask him to cut the glass 1/8'' scant on *each* side. There is nothing quite so frustrating as trying to fit an oversize piece of glass into a window frame that refuses to yield an inch—or an eighth of an inch.

The next step is to paint the inside of the window frame where the glass is to go. This is necessary for if you don't, the raw wood will absorb the oil in the putty leading to early putty failure—the putty will crack, even after you have painted it.

So, paint the window frame. Any grade of exterior paint is fine—the color is immaterial. Let it dry overnight. The next step is to make a bed of putty all around the frame for the glass. This putty bed need not be too thick, just enough so the glass will be pressing against the putty when it is placed in the window frame. Next, insert glazier's points or small brads to secure the glass within the frame. Now you can apply the putty—or window glazing compound, as it is now known. (Nothing like giving a fancy name to a common product!)

Roll the putty into a rope-like strand and push it in firmly with your fingers along the window frame and against the glass. Using a putty knife, held at a 45° angle, trim away the excess putty. Keep the knife moving in a downward direction. Try to finish one complete side of the window without stopping. Proceed in a similar fashion with the other sides of the window frame.

How about steel or aluminum casement windows? It is even more important with metal windows that the glass have adequate clearance between itself and the metal window frame. A close fit between the glass and the metal frame will result in a cracked pane. This crack will always occur at one of the corners of the glass because metal and glass expand at different rates.

If you are installing the glass in a metal window frame you cannot use glazier's points as the points cannot be driven into the metal sash. Use the special spring-like metal clips made for casements. Use one at each side of the frame. They engage the metal frame through holes drilled in the frame especially made for that purpose. After the glass has been "bedded" and the clips installed, apply the putty and trim away the excess, same as with wood sash.

The job is not complete until you have painted the raw putty. Best to do this after the putty has weathered a few days. Use a sash brush for this job and make certain that the paint makes a seal between the glass and the putty, otherwise the rain and sun are bound to lift the putty and crumble it away.

Glassware

Have you ever chipped an expensive piece of glassware? Possibly because you were in a hurry to wash the dishes and visit a neighbor or go to the movies—or you were just careless. There is no need to discard the item, especially if it is part of a set—and costly as well. Place the damaged glassware in an oven so the chipped area faces the front and will be easily accessible when the door is opened. Heat the oven to 450° F or so and let the glass remain in the oven for about 20 minutes. Light the propane gas torch. Open the oven door without disturbing the damaged glassware, play the flame of the torch on and around the damaged area until it starts to glow red hot. Soon the glass will start to melt slightly, "healing" the chipped area.

Turn off the oven and let the glass cool before removing it. When you do, you will be pleasantly surprised by the smooth edge the once-chipped glass now exhibits. This process will not restore the original contours of the glass, but it will be a vast improvement—after all, who wants to drink out of a goblet with a chipped edge?

Bending Glass

The propane gas torch can be used to bend glass tubing; for

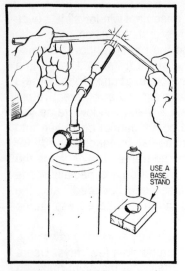

USE A
BASE
STAND

Bending glass tubing to make a straw is easy. You will need two hands so support the torch in a wooden stand as shown in drawing.

FINISHED
EDGE "MELTS"
ROUND

Raw edges on glass shelves can be a hazard. Round the edges by applying the flame of the torch to the edges until the glass melts.

example, to make a glass ''straw'' for drinking purposes or for making laboratory equipment. Light the torch and place it on a firm support so it will not wobble, as both hands will be used to hold the work. Hold the tubing in front of the flame and keep turning it until all parts of the proposed bending area are glowing red hot. When this occurs, start bending the glass tubing into the desired angle.

When the desired angle of bend has been reached, remove the tubing from the flame and let it cool. The result—a glass straw bent to the exact angle for imbibing fluids.

This is a great way to make children drink their milk as the novelty of seeing the milk appear in the tube somehow enhances the palatability of the milk. Why? We don't know—but kids are funny, as any parent will tell you.

Fire Polishing

The torch can also be used to ''fire polish'' the edges of a glass sheet that has been cut with a glass cutter. Just play the flame along the edges of the glass until it is red hot, keep moving the flame slowly so as to make sure heating will be uniform. Safety note: It is always a good idea to wear safety goggles when working with glass and use gloves if the heated part of the glass will be close to your hands.

Chapter 10 Working With Thin-Wall Conduits

Conduit, that thin-wall piping used by electricians, can be used to make many household articles—and probably the easiest item you can make is a simple step stool. Thin-wall conduit is inexpensive; a 10' length costs about $1.25 and is available from any large hardware store or electrical supply store.

To make the bends in the conduit you will need what is known as a hickey—a slang term for a conduit bender. You can rent one, but you may as well buy one—they cost about $5 and are usually sold without the handle. The handle consists of a length of 1/2'' pipe which is screwed into the head of the hickey. The hickey will enable you to make bends with a radius as tight as 5''.

The step stool illustrated has its platform 12'' from the ground. The first step is to bend the 2 U-shaped legs. Bend them into the shape shown with the help of the hickey. Because the hickey is not designed to make a bend greater than 90°, a little fudging will be required. To make the bend, keep slipping the hickey along the conduit and then bend the conduit slightly so it will pass the hickey handle. Of course, this operation will put a definite twist on the proposed legs. But no matter. You can easily straighten out the unwanted bend by inserting one end of the leg in a vise. Bend the other leg back into the position it should occupy. Bend a little past the center point as the conduit has a bit of spring in it. Check the success of your operation by placing the conduit on a flat surface. Both legs should be in contact with the flat surface.

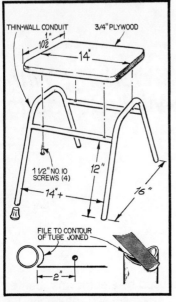

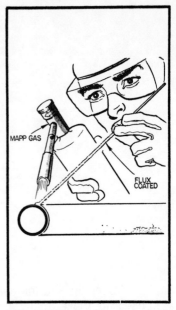

Step stool details. By upholstering the top, the stool will serve as a handy seat for watching television. Use polyurethane and Naugahyde.

With a little practice you should be able to braze the half-round horizontal members to the cylindrical uprights; wear goggles.

You will need 2 of these legs. So, by the time you get around to making the second leg, you should have benefitted by your experience in making the first leg. In addition to the 2 legs, you will need 2 crossbars. The crossbars should be about 15'' long. They are brazed to the legs at a point about 12'' from the ground.

A secret of a good brazing or welding job is a good mechanical fit between the parts to be joined. So, inasmuch as the tubing has a convex surface and the cut ends of the crossbars are flat, the next step is to file the ends of the crossbars into a concave shape to match the convex curve of the legs at the point of juncture. Place the crossbars in a vise and use a half-round file to do the shaping. Test frequently.

The next step is to remove the coating from the legs and the crossbars around the area where the joints are to be made. Do a conscientious job here as any zinc, dirt, grease or rust will adversely affect the brazing operation.

But before you begin, make a simple jig out of a scrap of plywood to hold the parts in alignment and to prevent movement. Clamps and weights can also be used to hold the parts in place. But whatever method you use, some sort of holding device is necessary to prevent movement during the brazing operation.

The BernzOmatic Super Torch can be used for this operation. The white, flux-coated bronze brazing rod can be used for the brazing operation. Heat the area to be joined to a cherry red, then apply the rod. It should flow freely upon contact with the heated metal.

The same business applies to other joints. You will have to make 4 all told. At the conclusion of the fourth joint you should be quite proficient as all the joints require the same technique. After all the joints have cooled, brush off any remaining flux with a wire brush and smooth the joints with sandpaper.

Next is the step. You can make it out of a piece of 1/2" or 3/4" plywood. But inasmuch as you have invested some time and effort in this project so far, you may as well devote some time and talent to making the step something a youngster can sit on in comfort as he glues his eyes to the television set.

Cut the piece of plywood so that it will be a loose fit between the sides of the legs and so that it will overlap the front and back by an inch or so. Visit an upholstery store, or a department store, and purchase a piece of 1"-thick polyurethane and some Naugahyde. If you paint the legs and crossbars black, orange Naugahyde will really make an elegant sitting stool.

Place the polyurethane over the plywood seat (it should be the same dimension as the plywood) and wrap the Naugahyde over the polyurethane. Tuck the edges over and around all sides and secure it by tacking. Don't be afraid to stretch the covering to make a neat cover. If you really are fussy, cut an extra piece of Naugahyde for the bottom of the underside of the seat. Fold the edges back and nail it in place with upholsterer's nails, the kind that have half-round or dimpled heads. This will effectively hide the carpet tacks on the underside of the seat. And if your tastes change, all you need to do is recover the seat with a new material.

We are nearly home-free in the construction of the step stool. Drill 4 holes in the cross bars 2 inches from the side of the legs. Set the seat in place and drill pilot holes for the screws. Be careful not to drill too deep—halfway through the plywood seat is enough. Now secure the seat in place with 4 No. 8 roundhead screws and the job is done. A set of 4 rubber or plastic crutch tips for the legs completes the job.

Thin-wall conduit can be used in the construction of many household, office and shop items. For example, it lends itself beautifully to the construction of a utilitarian bookcase. Now that

This larger version of the step stool is made mobile by the addition of casters at each leg. Note how a jig is used for the welding.

you have developed some skill in brazing conduits, try your hand at this slightly ambitious project.

As you can see from the drawing, a few more brazing operations are required than with the construction of the step stool. Limit the width of the bookcase to 36'' and its height to 5'. The shelves can consist of 1/2'' plywood, chipboard or pine. Before starting the brazing operation, check the height of the tallest book you intend to store in the bookcase and allow an extra 2'' for clearance (part will be taken up with the thickness of the shelf).

If the books or the items you are planning to store in this bookcase are especially heavy, a reinforcing bar brazed to the middle of each shelf support is a good idea.

The bookcase can be made mobile, no matter how heavily it is loaded, by inserting casters in the bottom of each leg. Any large hardware store can supply you with casters that are designed to fit inside hollow tubing. Get the kind with rubber wheels if the bookcase is to be rolled over a concrete floor; plastic wheels if the floor is carpeted. If the bookcase is to be stationary, use crutch tips to finish off the bottoms of the legs.

Chapter 11

Working With Wrought Iron

The BernzOmatic Oxygen Torch when used in combination with oxygen and Mapp gas makes an ideal tool for light welding projects made out of wrought iron and strap iron. Both of these metals are comparatively soft and can be made into many useful objects by welding.

Lowboy Plant Stand

Strap iron in 1/8" thickness and 1/2" in width can be used to make this simple plant stand. As you can see from the drawing, it is a 3-legged stand designed to hold a fairly large flower pot container 6" from the ground. The 3-legged arrangement makes it very stable and its height provides plenty of ventilation so that it is safe to keep the stand even on a carpeted area. (A plant stand in physical contact with carpeting or wood is bound in time to rot the carpeting and warp the wood floor.)

The first step is to make the 2 hoops for the stand, but before you do so, check the outside dimensions of the flower pot container that you are planning to reward with this stand. The inner diameter of the upper hoop should be 1/2" narrower in diameter than the outside diameter of the container. This is necessary so that when the stand is completed the upper part of the container will be a snug fit against the upper hoop. It will support the container and of course prevent it from falling through.

The same business applies to the lower hoop, but in this case the dimension is not too critical as the upper hoop is doing

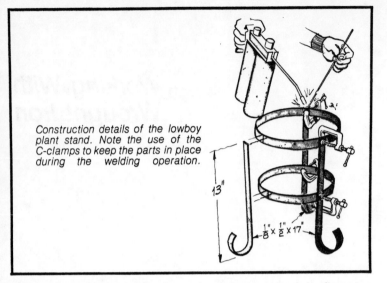

Construction details of the lowboy plant stand. Note the use of the C-clamps to keep the parts in place during the welding operation.

13"

$\frac{1}{8}$" × $\frac{1}{2}$" × 17"

all the work. But make the lower hoop large enough in diameter so that the bottom of the flower pot container passes easily through it.

Bend the 2 hoops into as near perfect circles as you can, clamp them in place with C-clamp and scrap metal and use the torch to weld the 2 ends of each hoop together. If the circles are not as perfect as Michelangelo would have made them, don't worry; a little judicious squeezing here and there will bring them into shape.

After the hoops have been completed, direct your attention to the required legs—for that's all it takes to make this stand: 3 legs and 2 hoops. Each leg is 17" long, however, inasmuch as each leg has a curved foot which takes up about 4 inches of metal, the finished legs will be only 13" long.

To make the curve at the end of each leg, heat the metal with the torch until it is red hot (no need for oxygen in this operation) and then make the bend with pliers. Again, don't worry if the curve is not perfect; after the metal has cooled, a little bending here and there will bring the legs into shape—after all, this will be a hand-made plant stand, not made by an impersonal machine, and some irregularity is permissible. (The ancient Persians, incidentally, *always* wove a minor imperfection into their rugs. They reasoned that only Allah can make a perfect rug and they had no inclination to challenge his workmanship!)

Now that you have completed the legs and hoops, the next step is to assemble the stand. The best way to do this is to clamp the legs to the hoops at 3 equidistant points around the periphery of the hoops in the position they will finally occupy.

Again use the BernzOmatic torch with the Mapp-oxygen combination to spot weld the legs in place. Two welds are required for each leg, one at the upper hoop and another joint at the lower hoop.

Wire-brush and sand the weld areas until they are fairly smooth. Sand the rest of the stand and apply 2 coats of black enamel. When dry, it is safe to insert the flowerpot container into the stand.

Place it in an area that is bound to get the attention and admiration which every plant needs.

Highboy Plant Stand

Obviously, this is a variation of the plant stand previously described. But because it is much taller, this plant stand can be used as a decorating motif to soften the corner of a room, the end of a wall or in fact any place where an eye-catching plant will do your home proud.

Note that this plant stand has 3 hoops or rings instead of 2. The bottom hoop serves to reinforce the 3 legs. Make the 2 upper hoops fit the particular flower pot container you have in mind. Inasmuch as most flower pot containers have tapered bodies, similar to a flower pot, this stand, as well as the lowboy model, will accept more than one size of flower pot container.

First step, of course, is to make the upper hoop, again using 1/8" x 1/2" strap iron. If you're interested in an exact dimension, the diameter of the upper hoop of this particular stand measures 9" (inside diameter) and the lower hoop is 7" in diameter, while the lower ring is 3" in diameter.

Make up the 3 rings again using the BernzOmatic torch with the oxygen-Mapp gas combination. One way to get a fairly round circle is to bend the strap iron around a circular form. A gallon paint can can be used as the form for the upper hoop, and smaller cans, or any round, stiff object as forms for the lower hoops. Try a convenient round fence post for size.

Put the 3 hoops away and apply your attention now to the legs. The legs are made out of the same strap iron as the hoops. Note that each leg has a small curved area at the top and a larger curve at the bottom. You will need 3 38"-lengths of strap iron. After the two curves at each leg have been made, the resultant length will be 30" from floor to upper rim.

Use the BernzOmatic Oxygen Torch to soften the metal to

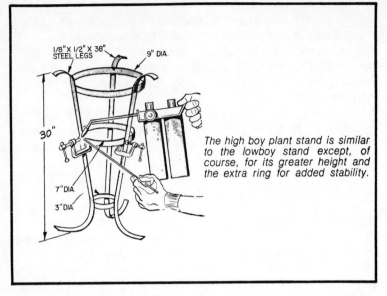

The high boy plant stand is similar to the lowboy stand except, of course, for its greater height and the extra ring for added stability.

make the curves for each leg. Note that the legs have a slight bend at their junction with the bottom hoop. This bend can be made after the stand is assembled.

To assemble the stand, weld the lower ring or hoop to the legs. Use clamps to make certain there is no movement during the welding operation. Next, weld the middle hoop in place—again using clamps. Finally, weld the upper ring in place.

As mentioned before, the outward bend of the legs from the lower hoop can now be made. Inasmuch as it is a rather shallow bend, it can be made cold. The angle of the bend will also determine the height of the stand as the greater the bend, the more the legs are spread out, thus lowering the total height of the stand. A height of 30" is about right.

Wire-brush the weld areas, sand thoroughly, apply 2 coats of black enamel or any other color if you have some special decorating scheme in mind. When dry, insert the plant holder complete with plant.

While we were in the process of assembling the stand, one of our neighbors dropped in to borrow a tool—since returned—and wanted to know what we were making. We told him it would be an umbrella stand which he believed until he saw the finished job complete with a philodendron plant. He has since stopped borrowing our tools!

Chapter 12 *Automobile Work*

If Americans have a love affair with their cars, then Saturday afternoons are the "date nights." That is the time when car tinkering starts in earnest from replacing a burned-out muffler to installing a new engine under the hood.

But let's start out with a fairly simple and all too common operation—removing a dent from the side of your sweetheart. Pounding from the inside of a door panel or fender just isn't the answer, as no matter how delicate a touch you may have, the hammer marks will still be visible—and then again, if the dent is on a door or body panel, you will have to remove trim and upholstery before you can start any pounding. But wait, modern-day science has come to your rescue with those ubiquitous epoxy compounds.

Get an auto body repair kit containing these remarkable epoxy compounds and finish off your repair job in an afternoon with the aid of a **propane gas** torch. Follow these steps:

　　1. Carefully clean the surface *around* the damaged area to remove all wax, body polishes, dirt and rust. This is important to prevent possible paint failure after the dent has been filled.

　　2. Clean and sand the damaged area.

　　3. If the dented area has a deep hollow and there is no way you can get to the back to force the dent back, drill a few holes in the metal and insert some sheet metal screws in the holes. Then place a bridge of hardwood over the

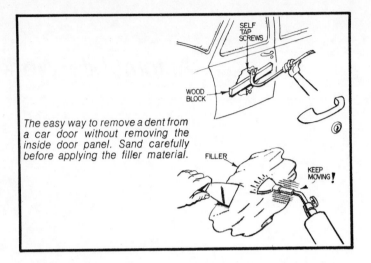

The easy way to remove a dent from a car door without removing the inside door panel. Sand carefully before applying the filler material.

dented area and use a claw hammer to pull the metal more or less back in place, using the wood as a support for the hammer. Pull gently, using slow, steady pressure.

4. Remove all the screws and carefully sand the surface with a coarse sandpaper to give a "tooth" to the body filler.

5. At this point the body filler should be prepared. It should be thoroughly mixed with the hardening compound that is part of the kit. Once mixed, the body filler has a limited life and should be used within a half hour. Ladle the body filler into the dented area, so that it is slightly higher than the surrounding area. At this point you will have a 24-hour wait to allow the epoxy body to cure. But you can hasten the curing process by *lightly* playing the flame of a propane gas torch over the body filler. Heat speeds up any chemical action and the hardening of an epoxy compound is a true chemical process—it has nothing to do with air-drying.

6. After the compound is hard—test it with your fingernail—mask off the adjacent area to protect it against any primer and paint overspray.

7. Apply the primer. When the primer is completely dry, sand with a fine wet-or-dry sandpaper; a No. 240 grit is recommended.

8. Wipe away any grit left by the sanding process and apply 2 or 3 coats of spray lacquer. Spray lacquer in aero-

sol cans is available in colors to match most American production-line cars.

9. The last step is a rubdown with paste wax and polishing with a terry cloth towel, and that's it.

Muffler Work

Back in the early days of motoring, mufflers lasted the life of the car. But nowadays, the extensive use of salts and chemicals to clear roads of ice wreaks havoc on auto mufflers—and yields a handsome profit to the muffler manufacturers and installers. But, if you are willing to get a little dirty by crawling under your car you can save two-thirds of the cost of a muffler installation by removing the old muffler and installing the new muffler yourself.

Check the Sears catalog for the muffler required for the year and model of your car. You may need some associated parts such as hangers and a tailpipe. If so, order them at the same time. When the parts arrive, check them to make sure they are for your car.

If you can get your car on a lift, fine. If not, you will have to improvise. Because present-day cars are built so low to the ground, some sort of clearance is required under the car before you can crawl under—a grim task if you are somewhat overweight! All you really need is about 4'' or 5'' of extra clearance.

You can get this by jacking up the 2 wheels on the muffler side and placing some timbers or bricks under each wheel. If you have a fancy car with dual mufflers, then work on one side of the car at a time. Make sure the timbers or bricks are wide enough to extend beyond the wheel area.

Before crawling under the car make sure the car is in park (or gear), the ignition key has been removed and the hand brake set. Now place a chock on *both* sides of the wheels resting on the ground. All these steps are necessary for your own protection, to make absolutely certain the car will not budge while you are operating under it.

Don some old clothes or a coverall and crawl under the car. Have your tools on hand and a BernzOmatic torch. The chances are that the nuts holding the muffler are rusted and no amount of tugging will budge them. Light the torch and, using the pencil flame, apply it to the stubborn nuts. A minute or less is usually enough. All you are trying to do is to expand the nut slightly to break its bond with the rusty bolt. If the bolts are still impossible to remove, and they often will be, use a BernzOmatic Oxygen Torch to cut them quickly. New bolts are supplied with your new

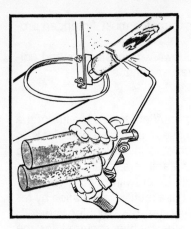

Flame-cutting is used as a last resort to remove stubborn parts. Caution: Wear goggles and be wary of molten metal falling on clothes.

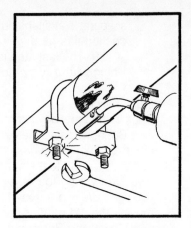

Those pesky, hard-to-remove tailpipe clamps can be made less recalcitrant with the application of a little heat. No sweat now!

muffler so you won't need the old ones. Be careful to direct the torch flame away from the gas tank and gas line and protect the area around the muffler from any sparks.

Even after all the nuts have been removed you may have some difficulty in removing the muffler from the car due to interference by chassis parts such as the propeller shaft, the rear end and the brake lines. In that case use a BernzOmatic Oxygen Torch to cut the muffler free. Do this cutting as close as possible to the muffler.

Before lighting the torch—and any other combustible materials—make sure that you will not be in the way of the molten metal as it drips down. If you are working in a garage, remove grease and old stains from the floor or cover them with sand.

Light the torch and place appropriate goggles over your eyes. Adjust the oxygen and fuel valves for a sharp edge to the orange flame and a sharp point to the blue; the blue flame should be about 1/4" in length. Apply the point of the blue flame to the area to be cut until it is yellow-white in color. When this occurs, turn off the fuel supply. Oxygen is all that is needed to keep the metal burning and cutting. If the metal stops cutting you are moving the torch too rapidly or else the oxygen cylinder is empty. You are working with extremely high temperatures, so proceed with caution.

Installing the muffler should be easier than removing the old one. After all, you are now working with a clean, rust-free muffler. A little judicious pulling and pushing should enable you to insert the muffler into its ordained position under the chassis.

One automotive accessory that could come in mighty handy in a muffler replacement job is a product called flexible exhaust tubing. This material sells for about $1 a foot and comes in many sizes to fit practically any exhaust system made. So, if a problem does arise and while under the car you discover that the tailpipe needs replacement and you didn't order a new tailpipe, visit an auto accessory store and buy a length of this tubing.

Measure the *outside* diameter of the tailpipe it is to replace and buy the tubing with the same *inside* diameter. To install the tubing, hold it in one hand and use the other hand to twist it. This will expand the tubing, usually just enough so that it will fit over other tubing or pipe that has the same external diameter as the internal diameter of the flexible tubing. You can tell which way to do the twisting, as twisting it the wrong way will decrease its diameter.

This tubing is easy to work with as its extreme flexibility allows it to pass over obstructions under the car by merely bending it and looping it over or around any hindrances.

Frozen Car Door Lock?
Yes, it happens, especially after a heavy rain followed by freezing weather. The usual recommendation is to install the key as far as possible and then to heat it in order to thaw out the lock. But how can you hold a short, hot key in your hand without danger of a burn? No way. A better idea is to use a heavy length of

Flexible exhaust tubing can be of great help when replacing a muffler or a tailpipe. It comes in sizes that telescope into each other.

The sensible way to open up a frozen car door lock. Insert a paper clip in the keyhole and then apply heat to the clip—not the lock.

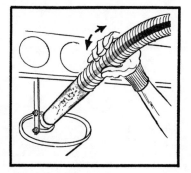
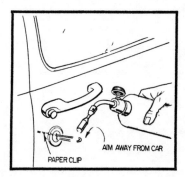

AIM AWAY FROM CAR

PAPER CLIP

wire, doubled over if necessary (part of a wire coathanger is fine) insert one end into the keyway and then apply the flame of a propane gas torch to the wire a few inches from the keyway—after all, you don't want to scorch the paint on the car. And this business of heating the key may not only burn your fingers but it may also deform the key so even after the lock is thawed out the key may not fit.

Chapter 13 *The Art of Burning Wood*

No, not burning firewood, but burning wood as an arts and crafts form. You can make repetitive designs for wall plaques, simple designs for salt and pepper shakers and imaginative trays for table use. All of these items, and countless more, will be true one-of-a-kind art forms—nothing that can be store bought.

Such items of wood, decorated in your own style, make gifts that are really personal, especially if the recipient's name or initials are burned into the gift.

In addition to a propane gas torch, you will need the following items: gloves, an asbestos pad, some steel wool and sandpaper, a hammer and a cold chisel, tin snips, a fine-tooth coping saw or a jeweler's saw, an awl, a steel brush, a spark lighter (part of the propane kit), some sheet metal, a brush and a can of wood varnish—and your fertile imagination. That's it!

Coasters
Let's start off with something easy to make—a set of 6 coasters. They will make a fine gift for a favorite aunt who will now be bound to remember you in her will. Cut 6 squares measuring 3 1/2'' x 3 1/2'' out of 1/4'' plywood. Let's assume your aunt's first name is Helen and your arch plan is to make up 6 coasters with a burned initial H in the center of each coaster. If her name happens to be Agatha, or Dorothy, the next operation is a bit more tricky as you will soon see. At any rate, her name is Helen and the first step is to cut the initial H out of the sheet metal. The

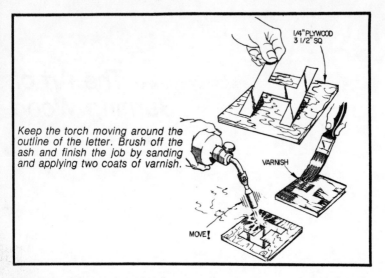

Keep the torch moving around the outline of the letter. Brush off the ash and finish the job by sanding and applying two coats of varnish.

1/4" PLYWOOD 3 1/2" SQ

VARNISH

MOVE!

sheet metal can be aluminum, copper, or even part of a tin can—as long as it is reasonably flat.

Use the awl to scratch out the H on the sheet metal and then use the tin snips to form the H. As we said before, if Agatha or Dorothy is to be honored, the A or the D will have to be cut so a "bridge" remains between parts of the letter, otherwise the upper part of the A and the inside of the D will be lost. Best to have an aunt whose first name is Helen, Laura, Susan, or similar first names that will not give you a problem in cutting out these respective initials!

The cut-out initial should, of course, be slightly smaller than the outside dimensions of the square—a 1/2" margin all around is a good rule to follow. Okay, you have cut out the required letter, make sure it is fairly flat—pound it lightly with the hammer and a block of wood if it isn't—and place the letter in the center of the coaster. Incidentally, the coaster can be round instead of square. Use the coping saw to make round coasters, or if you have access to a jig saw, use it. Much easier than cutting by hand.

Light the propane gas torch with the spark lighter and apply the flame, your talents and attention to the outline of the letter. Run the flame around the periphery of the letter, scorching the wood lightly as you do so. Remember, you are just trying to make an outline of the letter, not burn a hole in the wood, so don't overdo this part of the job. Be careful not to shift the letter while the burning is in progress. If you are satisfied with your "burnmanship," remove the letter from the wood with pliers or gloves. The letter H (if it is Helen) will now be plainly visible with a charred outline. Step back and admire your work. If all is well, proceed in the same way with the other 5 coasters.

Now that all coasters have been identified with the appropriate letter, sand the surfaces of all the coasters to remove any bits of loose burned wood. Blow away any ash that may be lurking in the crevices of the wood. Next step is to apply a coat of varnish to the coasters. Let the varnish dry overnight, sand lightly and apply a second coat of varnish. Let the second coat dry for at least 2 days and then apply a coat of paste wax, using fine steel wool as the pad. After a few minutes of drying time, polish with a terry cloth towel. The bottoms of the coasters can be covered with felt if there is any possibility of scratching—probably not—but this will show Aunt Helen what a thorough and careful craftsperson you are.

Caution: Use varnish, never shellac, to protect the surfaces of the coasters. A shellacked surface will show rings if a glass containing an alcoholic beverage is placed on it. While Aunt Helen may not drink, some of her friends may indulge.

A Cheese Board

Now that you have wet your feet—and hopefully not burned your fingers—in making coasters, how about trying something a bit more elaborate? A cheese board. This is not a board made of cheese, but a board on which cheese is cut—especially those hard cheeses which require a sharp knife for cutting and a strong constitution for eating.

First of all, get a solid board, not plywood, measuring about 10" x 12"—the exact dimensions are really not that important as cheeses come in many sizes and, after all, all we are interested in is a surface for cutting. If you have a circular saw, trim the edges at a 45° angle so that the bottom will be somewhat larger than the top. Not absolutely necessary, but it does add a bit of style to the project.

Now then, for our burning pattern. This time we are going to try something different. Instead of using the torch to burn *around* an outline, we are going to use the torch to burn an *interior* design. And our design is fairly simple—just a five-pointed star. The kind that appears on our flag—and for some perverse reason is also used by the Soviet Government as part of their heraldry.

Inasmuch as the tin snip will not be of much use in cutting small interior sections, use the cold chisel and a hammer to cut

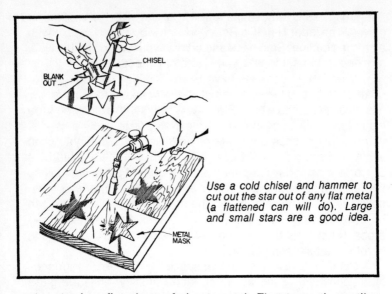

Use a cold chisel and hammer to cut out the star out of any flat metal (a flattened can will do). Large and small stars are a good idea.

out a star in a flat piece of sheet metal. First trace the outline with the awl and then proceed with the chisel and hammer business. What we are aiming to do is to burn a few random stars on our proposed cheese board. Mark the board in 5 or 6 places—no more—that will comfortably accept the stars without crowding. The sheet metal used for the star pattern should be large enough to protect the wood from any flame spilling over from the star area.

Position the sheet metal with the star pattern over the first spot selected for a star, light the propane gas torch and apply the flame to the cut-out area of the star. Use the flame long enough to char the wood while at the same time moving the flame into all areas of the star. Pay particular attention to the corners of the star. Lift the sheet metal patterns with gloves or pliers and examine your handiwork. If you are satisfied with the result—you should be as you should have considerable practice by now—proceed to the next star and the next star and the next until the full complement of stars has been burned into the board.

Go over the entire board with fine sandpaper making sure that no splintered or rough edges are evident. The final finishing job consists of an application of pharmaceutical-grade mineral oil to seal the pores of the wood. Do not use an organic oil such as cottonseed oil or peanut oil as these oils tend to pick up an odor and get rancid in time. Caution: Do not seal the wood with lacquer, shellac or varnish, as the cutting operations while slicing the cheese will only break the film.

The board of course can also be used for other kitchen chores such as slicing tomatoes, cucumbers and bread—but please, no onions, as they will be sure to leave their distinctive odor—and who wants cheese flavored with onions?

Note that in the above operation the flame of the torch was applied to the open areas of the sheet metal pattern. You can just as well reverse this operation by making the stars the solid part of the pattern and applying the flame to the periphery of the star pattern—which of course will yield a star, but this time the charred area will *outline* the shape of the star. And there is no need to stick with stars. Crescents, fleurs-de-lis, flowers and even animals can be used as the design motif.

Wall Plaques

Wall plaques are a sort of giant-sized version of a cheese board. But inasmuch as a large piece of wood should be used for a plaque, say 16'' x 20'', the design should be bold and big. For a child's room an abstract pattern of Jack-on-the-Beanstalk, Little

Peter Rabbit can do double duty. Lightly scorch the wood around the cut-out figure. Or you can also use the "waste" and scorch the inside.

Red Riding Hood, Bambi the deer, Peter Rabbit, or Snow White are very suitable.

Copy the design from one of the children's books, which are generally pretty well tattered and can miss a page in the interests of art. Use carbon paper to transfer the design to the sheet metal. Make "bridges" at open areas so they will not fall through when the cutting operation is started. Cut out the pattern with the tin snips and with the help of the cold chisel and hammer. If you are careful in this operation, you may be able to wind up with 2 patterns, one a "negative" of the original design. Either one can be used for the burning operation and in fact if you make up a pair and mount them side by side, you will have a sort of "night" and "day" set of plaques.

A Treasure Chest

And now for something a bit more ambitious. A treasure chest that can be used to store a child's toys, winter garments or just miscellaneous knickknacks that are just "too good to throw away."

The chest is made out of inexpensive pine wood or fir plywood. It is 48" long, 15" wide and 18" high. And in fact it can double as a bench for sitting when placed against a wall or as a window seat. The cover overlaps an inch on all sides and has a catch so that it will stay in an open position for "loading" or "unloading." Use 3 hinges for the cover, or a long piano-type hinge to secure the cover to the box proper.

All the joints are simple butt joints, put together with white glue and nails. Check carefully before driving the first nail to make sure the joint is at an exact right angle to its neighbor.

To make the chest mobile, mount 4 casters at the bottom. If the room where the chest is to be located is carpeted, use casters with hard plastic wheels; if the room has a hard surface such as a hardwood floor or tiles, then use casters equipped with soft rubber wheels.

The next step involves the use of a propane gas torch to give the chest a truly distinctive look. Before lighting the torch there are a few facts you should know about wood which will enable you to do a better job of wood burning.

A tree, as most of us know, indicates its age in years by its annular rings. The narrow rings represent the winter growth and the wider rings are the summer growth. These wider rings are much softer in texture than the narrow, winter-growth rings. It is this difference in texture that allows us to use the torch as a decorative tool. The soft summer growth will scorch and burn much more rapidly than the tougher winter growth. Of course the rings, per se, are not apparent on boards or plywood, but the sawing operation in the mill exposes hard and soft sections of wood which at one time corresponded to the winter and summer growth rings.

Now that we have finished our lesson in forestry it is time to begin. Light the propane gas torch and play it lightly over a small area until the wood starts to char. Continue to an adjacent section. You will soon notice that some parts of the wood char much more readily than other areas. At this point you can commend yourself upon your powers of observation! The soft summer growth does char much more rapidly than the hard winter growth.

Construction details of the child's chest. After assembly has been completed, use the torch as shown to burn away some of the soft wood.

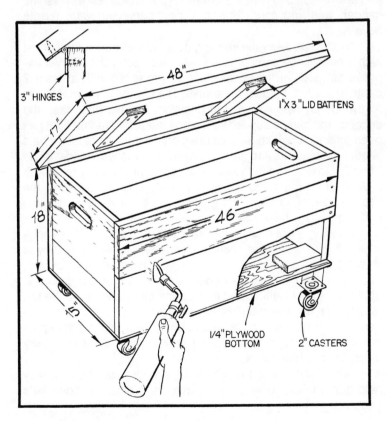

Continue in this fashion over the front, top and sides of the chest. There is no need to treat the back of the chest unless you are a perfectionist who insists upon finished craftsmanship even when not visible to casual inspection. Good, we like that kind!

There is no need to char every single area of soft growth. Apply the torch so that you will get an overall effect that will be pleasing to the eye—don't overdo it. You don't want to wind up with a chest that looks like a giant piece of burnt toast.

After burning and charring have been done to your satisfaction, the next step calls for an application of the well-known elbow grease. Use some fine sandpaper and go over the entire chest until it feels as smooth as silk. The unburnt areas should be sanded just as carefully as the burnt areas. After you have gone over the entire chest—and you are satisfied with your work —go over it again, this time using a finer grade of sandpaper and always sanding with the grain. If you have a power finishing sander, well and good, use it; if not, persevere with muscle power.

Next, wipe away the sanding dust from all parts of the chest with a rag slightly dampened with turps or paint thinner. All exposed surfaces of the chest must be protected with at least 2 coats of shellac, lacquer or varnish. The choice is yours, but the following may help you in your decision. Shellac dries fast, the second coat can be applied within an hour after the first coat has been brushed on; lacquer also dries quickly, but at least 3 coats should be used as lacquer, sprayed or brushed, tends to be "thin"; varnish is an old reliable, 2 coats will do for varnish with a 24-hour drying period between coats. Before applying the second coat of varnish, lightly sand the surface, wipe it clean and then apply the varnish.

Sanding is not really necessary between coats when applying shellac or lacquer as shellac and lacquer have a tendency to dissolve into the preceding coat—which makes it easy for you.

After second coats have been applied wait a few days and apply paste wax well rubbed in with fine steel wool to all the finished surfaces. The steel wool will cut and remove the dust particles which inevitably have settled on the work during the application and drying periods while the wax will yield a fine, silky finish to your handiwork.

And now it is time to present the chest to your child. It may help him, or her, to keep that playroom a bit more tidy—let's hope so.

Book Shelves

The propane gas torch can be used most advantageously to decorate and give character to many other items around the house. For example, how about those wooden boards used for book shelves? Pretty plain looking, right? Well you can really doll them up by burning the edges of the shelves. Just play the flame of the torch slowly along the front edge until the degree of charring meets with your esthetic sense. Wipe the charred area with a damp cloth. There is really no need to do any finishing work on this particular job as the edges of the shelves are not subject to any noticeable wear.

Picture Frames

Those drab, plain wood picture frames can be given a new lease on life with the help of a propane gas torch. Remove the glass, mount and picture, and lightly scorch the 4 sides of the frame. A light sanding and an application of lacquer or shellac, rubbed down with steel wool and wax when dry, finishes the project. Picture frames treated in this fashion go especially well with colonial-type furnishings. This same technique can also be applied to mirror frames.

No need to put up with plain wooden shelving when a little time and effort—plus the torch—can make the plainest of shelves decorative.

Picture frames of wood are a cinch to dress up. Play the flame of the torch around the wood until you get the effect you want. Keep moving!

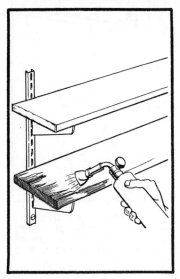

Switch Plates

If you would like to see something different in the way of switch plates for your home, consider switch plates—and outlet plates—made out of wood. Fairly thin wood should be used, never more than 1/4" thick. Cut the wood to the same size as the original plates to be replaced. If you have a power table saw, cut the wood so it will have a 30° bevel all around. Once the saw is set up, you can cut several at a time without changing the setting of the fence or the tilt of the saw blade. Play the flame of the propane gas torch around the edges of the switch plates, cut the appropriate holes for the mounting screws and the toggle, sand, finish with lacquer, shellac or varnish and the job is done.

Drawers

Now here is a place where the most mundane chest of drawers can be made to shine and sparkle with character. Remove each drawer and lightly burn the outside edges, making a charred area of about 1 1/2" in width. Sand and finish in the usual way and replace in the chest. This is one time it will really pay you to step back and admire your results for they are well worth admiring. Of course this process can only be applied to drawers that have not previously been painted.

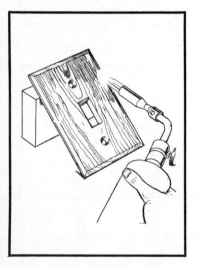

How to really individualize the switch plates in your home. No two plates will be the same. Use fairly thin wood; bevel all edges.

Even an old dresser can be given a new lease on life by lightly burning all four edges of each drawer. Pull drawers out as shown here.

Use the wide-flame attachment on the torch to char the edges of a door that was made out of planks and battens. Finish by varnishing.

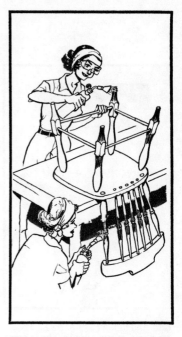

Plain, old, wooden kitchen chairs can be made to look like a professional decorator's pet by charring the bottom of the four chair legs.

Doors

Doors can be similarly treated. Again this applies to unpainted natural-finish doors. This time the charred area should be somewhat wider, about 2'' in width. Such treatment works best with doors that are part of a rustic cabin, a summer home or doors leading to a room with a Colonial or Early American motif.

Chairs

Those inexpensive natural-finish maple chairs, usually called factory chairs because of their popularity in the garment-trade sewing centers, make ideal candidates for dressing up with the aid of the propane gas torch. Turn the chairs upside down and lightly scorch about 4'' of the bottom section of each leg. Then apply the torch to the middle of the back, and to the middle of each rung—no more, please. A light sanding and finishing should complete the job.

You will now have a chair—or a set of chairs if you do more than one—that can really be a decorator's delight, the kind of

Charring the outside of a salad bowl for some mysterious reason makes the salad tastier. My grandmother says it's some form of osmosis.

chairs that visitors will be sure to say, "Where did you get those elegant chairs?" You can answer with pride, "I made them myself." And so you did—or at least the decorating part.

Salad Bowls
Those large wooden salad bowls can be made to look a bit more exciting than a plain, old wood salad bowl with a little heat treatment. Run the **propane gas** torch in an up-and-down curve along the outside of the bowl. Your guests will be sure to admire your handiwork and may even elicit a favorable word about the salad inside the bowl. The same treatment can be applied to those ubiquitous little bowls that always accompany the large bowl as part of the purchase. While the outside of these bowls, big and little, may be varnished (the insides seldom are), this will not affect the charring or the design. Sand and finish as described before.

Kitchen Cabinets
Natural-finish kitchen floor and wall cabinets made of birch, oak,

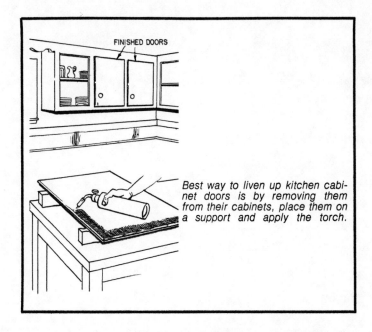

FINISHED DOORS

Best way to liven up kitchen cabinet doors is by removing them from their cabinets, place them on a support and apply the torch.

maple and pine lend themselves especially to treatment by charring. The best way to do this job is to remove the doors from the hinges. Bring them down into the basement and place the doors over a box—or sawhorses—so that the edges will be at a comfortable working height. Play the flame of the torch along the 4 edges of each door to produce the desired effect. Proceed in a similar manner with the balance of the doors until all the good work has been done. Sand and finish in the usual manner and rehang the doors.

Other Items
Even inexpensive souvenir items such as ash trays, salt and pepper shakers and similar knickknacks can be highly personalized with the use of a **propane gas** torch. An initial, a star, a flower, a daisy pattern and other designs can be burned into these objects. Your own imagination is your limit and if you have read so far, your imagination at this point should know no bounds!

If you are planning to go into any aspect of jewelry making involving metals, then you should become thoroughly familiar with the technique of hard soldering. As indicated in Chapter 1, hard soldering differs in many respects from soft soldering. In fact, it is a close cousin of brazing. Let's examine hard soldering in greater detail.

First, something about silver, as silver is an important component of many so-called hard solders. Sterling silver has a melting point of 1,640°F. By tradition, sterling silver is composed of 925 parts of pure silver and 25 parts of copper. By comparison, coins made of silver (we are not referring to the debased coins made of copper and clad with silver—dimes, quarters and half-dollars) have 900 parts of silver and 100 parts of copper.

Hard solders for working with silver (and other metals) have melting points ranging from 1,200° to 1,450°F. Note that all of these temperatures are well below the melting point of silver. This is most important as a solder whose melting point is too close to the metal to be worked on may result in the parent metal melting along with the solder.

Using solders with varying melting points is most important in many phases of jewelry making. For example, a high melting point solder can be used for initial soldering and then lower temperature solder can be used for subsequent operations. This way there is no danger of melting the solder used in the first application.

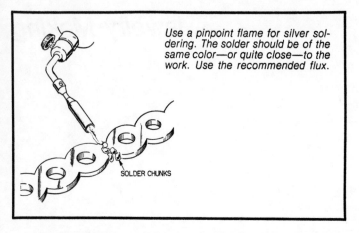

Use a pinpoint flame for silver soldering. The solder should be of the same color—or quite close—to the work. Use the recommended flux.

SOLDER CHUNKS

Hard solder is available in wire form, in sheets and even in tiny 1/16" squares. For the best all-around use, buy hard solder in sheet form for jewelry work.

The Flux

In addition to the solder you will need a suitable flux. (The word "flux" is derived from Latin, meaning "to flow.") A flux commonly used with hard solder is borax. This material is mixed with water until it has the consistency of cream. Boric acid is sometimes added to a borax flux in order to raise the point at which the flux will become a fluid under the heat of the propane torch. In fact, you can get fluxes which become fluid at definite ranges, thus giving the user a range of temperatures, important when a series of soldering operations are necessary on the same piece of work.

The first step, and an important point always to bear in mind when doing hard soldering, is *cleanliness*. The parts to be joined must be absolutely clean of grease, dirt, rust, oxides and similar gremlins. Cleaning can be done by 1 of 2 methods: pickling or sanding. Silver and copper can be effectively cleaned by the pickling process. This consists of immersing the items in a pickling bath of 1 part of sulphuric acid to 10 parts of water. *Caution: Always add the acid to the water, never the other way around*. When copper is immersed in the pickling bath, it will turn a rosy pink; when silver is immersed, it still will come out a pure white in color. This is due to the fact that the acid in the pickling bath has dissolved all the copper in the silver—or at least on the surface.

The pickling bath can be used over and over again, until it no longer has any effect on the metal. But for occasional hard soldering, an abrasive method of removing surface impurities will work just as well. Filing, steel wool or sandpaper, plus muscle power, will do an effective job.

Fitting the Pieces Together

There is no point in expecting solder to span a space between 2 poorly fitting pieces. It is not in the nature of solder to be that accommodating. A good, close fit is very important when doing any soldering work. You see, hot solder tends to flow between adjacent pieces of metal, no matter how close the fit is, by a process the chemists call capillary action.

So what can you do if the 2 pieces you are planning to join simply do not fit snugly? If it is too costly or impractical to make the piece, or pieces, over, then the next best thing is to fill up the gap with small bits of the same metal, then proceed with the soldering business. Any excess metal that remains after the soldering has been completed can be filed away. No one but you will know.

Applying the Flux

After the surfaces of the embryo piece of jewelry have been cleaned—and we mean cleaned just prior to this operation, not yesterday or even a few hours ago—apply the flux to the joint area with a brush. Use the brush to pick up a few bits of solder— if you are using the flat type—and place it over the fluxed area. During the heating operation, additional solder may be required.

Obviously, a source of heat will be required for soldering and a propane gas torch is ideal for this step. Its heat is constant (unless you run out of fuel!) and is easily controlled.

Color can be a good indication of the temperature of the work as the torch is applied. Shade the work from extraneous light such as sunlight or an overhead lamp. When the work first starts to turn a visible red, then the temperature is around 900°F. An over-all red hue means that its temperature is 1,200° while a bright cherry red indicates a temperature of 1,400°. When the work starts to turn pink, back off. This is the danger point; it means that the work is flirting with possible melting. It has reached a temperature in the neighborhood of 1,600° and don't forget that silver melts at 1,640°F.

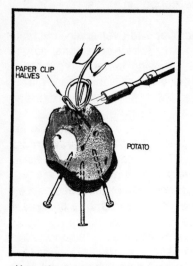

PAPER CLIP HALVES

POTATO

How a lowly potato can be used to support a high-quality gem. Note the use of the paper clips to steady the stone, nails serve as the legs.

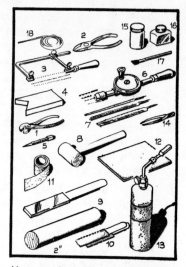

Here are the tools and accessories required for making jewelry. The investment is a small one. Always purchase tools of highest quality.

A good technique to follow in silver soldering jewelry is to preheat the work. Play the torch over the entire surface of the work and when it just starts to glow red, concentrate at the joint area. The solder will always flow to the hottest part of the work—provided you have carefully cleaned the joint area. So, if heat is concentrated on the joint, the hot metal will melt the solder—sort of doing the work for you. Never try to melt the solder first, it will only ball up and refuse to flow where you want it to go.

Caution: Always use a fire brick as a "table" for soldering work. An ordinary brick may contain trapped water and may explode under the intense heat of the torch.

Protecting Stones

Semi-precious stones and precious stones do not take kindly to heat and they should either be removed from their settings or protected from the heat of the torch. Best, of course, is to remove the stones; however, in many instances this may be impossible because they may form an integral part of the piece to be worked on. For example, if a break in a ring is to be repaired and the amethyst cannot be removed the work can still be done without harm to the stone. Visit the family larder and select a potato that is somewhat oval in shape. Jab 3 nails into the narrow end to form a sort of tripod. Make a small cross on the opposite end and insert the stone of the ring into this opening. Make sure the entire stone is embedded in the potato. Immobilize the ring during the subsequent soldering operation. Cut a paper clip in half, saving the U-shaped sections. Use these halves as a further support for the ring, placing 1 at each side of the stone.

Use a small flame and work rapidly to avoid transferring too much heat to the stone. When the job is finished, not much damage has been done to the stone or to the spud—it is still edible.

Now that we have learned, more or less, the fundamentals of jewelry making, how about trying your hand at an actual piece of jewelry? This one is really going to be custom made and we are indebted to Linda Gibson of Adelphi College for her help.

Parts and tools for jewelry making can be purchased from any crafts and hobbies shop. Look in the Yellow Pages under Arts and Crafts Supplies.

You should have the following tools on hand:
1. 6'' diagonal pliers
2. Cutting pliers
3. 4'' jeweler's saw and several blades
4. Wood block (bench pin)
5. Nail or center punch
6. Hand drill and several small bits
7. Files: half-round, medium cut; several needle files; round and triangular files
8. Small mallet (rawhide or wooden)
9. Baseball or softball bat
10. Several felt polishing sticks
11. Emery paper or Aloxite cloth No. 320
12. A piece of hard, non-flaking asbestos board to protect the work surface
13. A **propane gas** torch
14. Metal tweezers
15. Small bottle Sparex (cleaning solution)
16. A bottle of silver flux (for keeping metal clean in soldering process)
17. Small brush for flux
18. Silver solder (in sheet or wire form)
    1/4-pound bar of Tripoli compound (for initial polishing)
    1/4-pound bar red rouge (for final polishing)

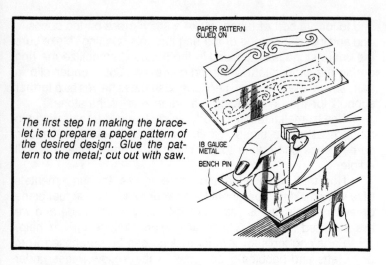

PAPER PATTERN
GLUED ON

18 GAUGE
METAL

BENCH PIN

The first step in making the bracelet is to prepare a paper pattern of the desired design. Glue the pattern to the metal; cut out with saw.

To make a bracelet, you will need the following materials:
1 piece No. 18 gauge sheet metal 2″ x 6″ (brass, bronze, copper, silver)
A 15″ length of No. 14 gauge square metal wire

1. After measuring the wrist size (yours or the recipient's) make a paper pattern to the exact measurement and in the exact width you want. Sketch the design on the paper. This can be fairly ornate, such as tracing a design from a picture; or fairly simple—initials, for example, within a stylized border. Use white glue and cement the pattern on the metal and let dry.

2. Place the work on a wood block (bench pin) for support. Holding the saw vertically, saw up and down in long strokes, cutting close to the edge of the design, but not directly on it. Next, using a nail or center punch, make a dent in the center of the design. With a hand or electric drill and small No. 58 bit, pierce through the dent. The hole is for the saw blade to slip into. Release an end of the blade from the saw frame, insert blade through the hole and then secure the back to its position on the frame. This procedure is done with all parts of the design, and while it is time-consuming, the results, as the design starts to appear, will more than repay the effort.

3. Using a half-round file, smooth all rough edges made by the saw. Work on the bench pin and slide the file on the bracelet edge in long, gliding strokes. File slowly until the desired edge is

achieved. Smooth the piece if necessary with a fine-cut file. For the small inside areas use the needle files. Take your time and work gently. When the filing is finished, run warm water over the bracelet so the paper can be removed. Dry and sand the surface with an emery cloth to remove every bit of glue.

4. Soldering may seem rather complex, but after the first time it will become quite easy. Since this step is most important to *all* metal jewelry making, work slowly and meticulously and you will be successful.

Clean, cut and shape the square wire so it fits snugly against the entire edge of the bracelet. Next, apply the flux to the bracelet with the small brush. Use the emery cloth to clean the solder and cut it into several dozen 1/16'' pieces. Pick them up with the brush and place them on the damp flux where the 2 metals will come into contact.

Light the torch and begin applying its heat to the solder until it flows. As the solder flows and bonds the 2 metals together, let the bracelet rest and cool.

The metal will likely look black at this point. To remove the effects of the heat oxidation, place the bracelet into a glass or plastic bowl which contains 1½ cups hot water and 1/4 cup Sparex. This solution will clean the metal. Rinse the bracelet in clear water and dry.

5. Using the rawhide or wooden mallet, gently hammer the flat bracelet into a round shape on the baseball bat. Remember,

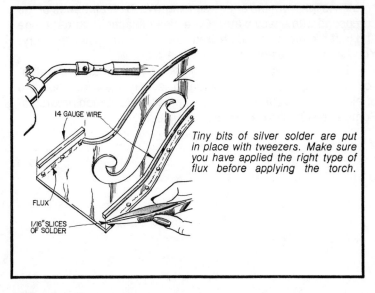

14 GAUGE WIRE

FLUX

1/16" SLICES OF SOLDER

Tiny bits of silver solder are put in place with tweezers. Make sure you have applied the right type of flux before applying the torch.

the bracelet must slip easily over the wrist, so leave sufficient room between the 2 ends.

6. You have reached the last step! The bracelet is now ready to be polished. It is easier to polish and buff with an electric buffing wheel, but felt sticks will achieve the same purpose with a little more effort. Rub some Tripoli compound onto the felt stick and rub it over the metal surface to remove any fine scratches left from the emery sanding. Wash off all traces of the compound with soapy water. On a clean felt stick rub red rouge and buff the metal to the desired sheen. Wash again, then dry. You have now designed and made a bracelet!

If making the bracelet has whetted your creative appetite, you may want to take a course in jewelry making. They are offered in Y's, in evening schools and adult education programs. Some excellent books are also available.

Chapter 15 *Enameling*

Top-grade enameled-copper pins, ashtrays, jewelry and even switch plates are well within your skill if you use a Bernz-Omatic torch as the source of heat. No need to spend hundreds of dollars for a kiln (incidentally, the word is pronounced *kill*, the *n* is silent).

You will need an oven—and all it consists of is a large juice can cut as shown in the drawing. The required shelf is cut out of "hardware cloth." Mount it as shown. The shelf supports the work during the heating process. All the can really does is to form a heat reflector so that none of the heat from the torch is lost. Support the can-oven on a couple of rods.

All of the rest of the parts needed can be purchased from hobby shops—look them up in the classified telephone directory. You will need copper in sheet form, a couple of jars of colored glass "dust," sprinkles or beads and a binder to keep the glass in place. Oil of lavender or gum tragacanth will do nicely as a binder. In fact, you can cut your investment in copper to 1¢— yes, a penny can be used for your initial project—it makes a very fine pin with the addition of a clasp on the back.

Let's start off with the penny project. Clean Lincoln's face thoroughly. Rub it on a medium-grade sandpaper until it shines like the proverbial new penny. Be careful in handling it as the oil from your fingers can affect the enameling process. (The pores in your fingers secrete an oil, no matter how recently you have washed your hands.)

Next, brush on a thin coat of the oil of lavender or the gum tragacanth. Now sprinkle on the colored glass. If you have 2 or more colors of the glass, you can make a pattern on the penny. Sift enough of the sprinkles on the coin so that the entire surface of the coin is covered. No copper should be visible; any excess, even though mixed, can be saved for future use. Use tweezers to place the penny and its glass sprinkles on the hardware cloth shelf.

Everything is set for your artistic talents. Light the propane gas torch and apply the flame to the *bottom* of the penny resting on the shelf. Wear goggles—just in case. Keep the torch in a horizontal position just slightly tipped up so that the flame can impinge on the penny. If you try to use the torch with the tip lower than the body of the torch the flame may go out.

Keep up the good work with the torch until the glass starts to melt. You have now established your artistic worth and if you want to improve on your maiden efforts, add a few more sprinkles of glass until the desired effect has been obtained. Wait until the penny has cooled before you handle it, or if you are really impatient, remove the penny from the oven-kiln with the tweezers and place it on a fireproof surface to cool. This way it will be easier to admire.

The pin—for that is what it will ultimately be—is finished off by cementing a catch or clasp to the back of the pin. These catches are known in the hobby trade as jewelry "findings" and are very reasonable in cost—$1 will get you a dozen assorted catches, loops and "rings." Use epoxy cement to fasten the catch to the back of the pin.

A Belt Buckle

How would you like to customize a belt buckle at no cost? It can be done. This time, instead of a penny and colored glass beads, a metal belt buckle and colored glass is used. The glass is for

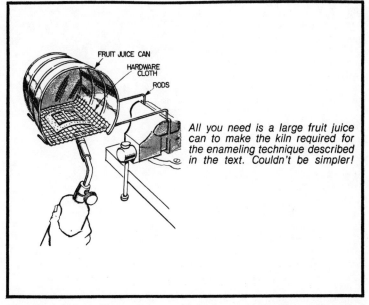

FRUIT JUICE CAN
HARDWARE CLOTH
RODS

All you need is a large fruit juice can to make the kiln required for the enameling technique described in the text. Couldn't be simpler!

free. Any colored beer, wine, whiskey or soda bottle can be used for the glass. Wear goggles and break up the glass bottles or jars into small bits, about as big as grains of barley. Thoroughly clean the front surface of the buckle, add the binder and place some of the glass bits over the buckle into the pattern that will please you most, or will be apt to please the recipient of the buckle. Light the torch and apply the flame to the bottom of the buckle until the glass melts. You can even poke the molten glass with a wire to get a particular design the artist in you dictates.

Ashtrays and Switch Plates
The same business can be applied to any metal ashtray and even to switch plates. Your own imagination is the only limit to what you can achieve. Try larger bits of glass to get big blobs of color. You can add the glass—beads, sprinkles or glass bits—during the firing cycle as the glass is melting if you want to change the design or pattern. Small bits of metal can also be used. Wait until the glass is molten and then use tweezers to drop in the metal. Keep it up and soon you will have enough items to have your own one-man show—or a one-woman show.

Chapter 16 *Repair Work*

"Ma, Jimmy broke the basketball basket!" And so he did. How? By showing his friend that he could chin more times than his friend. Well, the number of times Jimmy could have chinned is still in dispute as after the third try the hoop gave way and deposited Jimmy on the ground—unhurt but mad because his friend had chinned more times than he.

There is no need to rush out to buy a new basketball set-up. If you will examine the break carefully, you will note that the fracture occurred at the same place where the factory welded the hoop to the supporting bars. All you need do is apply your new-found skill as a welder to make the repair.

Clean and prepare the metal as described earlier and use the BernzOmatic Oxygen Torch to make the weld. Make sure the hoop is securely clamped in position and will not move during the welding operation. A bronze, flux-coated or nickel silver rod should be used as a filler as the weld is being made.

When the weld has cooled, clean the weld area with a wire brush and sandpaper. Paint the entire hoop assembly with black enamel. Forgive Jimmy.

Cultivators and Garden Rakes

These garden tools always seem to have a loose or missing tooth at the beginning of the gardening season. Their repair with the aid of a BernzOmatic Super Torch and Mapp gas or a BernzOmatic Oxygen Torch is a comparatively simple exercise in brazing.

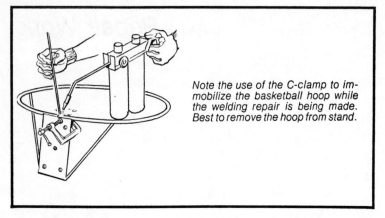

Note the use of the C-clamp to immobilize the basketball hoop while the welding repair is being made. Best to remove the hoop from stand.

Clean the areas around the break with a wire brush and coarse sandpaper, making sure that every last vestige of rust, dirt and paint has been removed. Support the work as shown in the drawing so there will be no possibility of movement during the brazing operation.

Inasmuch as all cultivators and rakes are made of steel, you can theoretically use any one of the 4 brazing rods recommended for steel brazing. However, as this is not a jewelry repair, there is no point in using silver solder. Instead, use the nickel-silver rod or the flux-coated bronze brazing rod.

Heat the parts to be joined until they are a cherry red and then apply the rod. It should flow freely in contact with the heated metal. Be careful not to move the parts during this operation. Wait until the joint has cooled before testing the efficiency of your handiwork. The chances are that you did a good job and you will now have no excuse to delay cultivating and raking your vegetable patch.

Frame Repairs

There are many household items built around a hollow tubular frame such as lawn mowers, snow blowers, hose racks, swing sets, motor bikes, bicycles, barbecue grills, children's play equipment and similar items. A break in the framework of these items is usually caused by sudden stress at an area that has been already weakened by rusting. Merely welding or brazing the parts together is at best a temporary repair as the rusted area almost always extends well beyond the break.

What to do? The defective areas should be reinforced as well as welded or brazed to make the repair. A practical way to do this is to "wrap" the area with steel tubing. You don't need much steel tubing to make such a repair. A 6" length is usually enough. Steel conduit, the kind used by electricians for wiring, is ideal for this purpose. They come in various diameters, depending upon the number of wires they are to carry. An electrical supply store can probably sell you some scrap lengths, enough to last you for many repair jobs. The exact diameter is not too important as the technique of making this repair calls for cutting the tubing lengthwise to make 2 equal sections about 6" or so long.

Make a "dry run" by placing each half of the tubing around the damaged frame. If the tubing fits loosely or is just a bit too tight for a good fit, use a hammer to "customize" the patch. Pound the edges if the fit is too loose and apply the hammer to the high part of the tubing to make it somewhat wider in diameter.

After you have achieved a reasonably good fit, thoroughly clean the insides of both halves of the tubing and the outside of the frame you are planning to repair. Use sandpaper, steel wool, a wire brush and plenty of muscle power to expose the bare metal. Then proceed to make the repair using a BernzOmatic Oxygen Torch if the repair is a particularly "heavy" job, or the BernzOmatic Super Torch if the work calls for a somewhat lighter application.

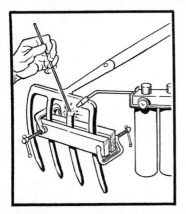

How a broken tine on a rake can be repaired by welding. The use of the wood clamping arrangement is a good idea that prevents loss of heat.

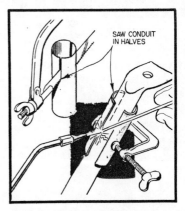

Ordinary steel conduit, used by electricians, slit in half as shown, can be used to repair many household items. Final step is welding.

Chapter 17 *Additional Uses for Your Propane Gas Torch*

Let's assume that you are on a camping trip and a cup of hot tea or soup would just hit the spot. No need to build a campfire or haul out your portable range. Use the torch. Water in a cup can be rapidly heated by directing the flame of the torch to the surface of the water as shown in the drawing. If the water or the soup is in a metal container, say the can in which the soup came, direct the flame of the torch to the lower half of the can. *Do not apply the flame to the bottom of the can.* This will necessitate tipping the torch in an upside-down position and the flame will be sure to sputter and go out.

As we all know from our high school physics, water is very efficient in absorbing heat from metal, which is why you can put your hand on the radiator of a car and it will feel just uncomfortably warm, even though the water inside the radiator is boiling. And it is for this reason you can bring a can of soup to a simmer without melting the metal of the can. *Caution: Place the can on a firm support, such as a flat rock; don't hold the can in your hand.*

Camping out? No need to break out the camp range just to heat up a small saucepan. Use the torch and heat the sides until food is warm.

Lemon meringue pie looking pale? A light toasting with the torch will allow your guests to tell you what a wonderful cook you are.

Gilding the Lily

Remember that beautiful lemon meringue pie your wife made? So what if it doesn't have that toasted top that looks so succulent in the color picture accompanying the recipe. And your wife slaved so hard over it—and company is due in 10 minutes. And your oven is otherwise engaged. Tell your wife not to fret. A few passes of the flame of a propane gas torch will put an elegant brown top to that pie. Pass the flame lightly over the pie top until the ''crust'' turns a golden brown. Don't overdo it—you are not trying to incinerate the pie, just trying to help your wife out of a trying predicament.

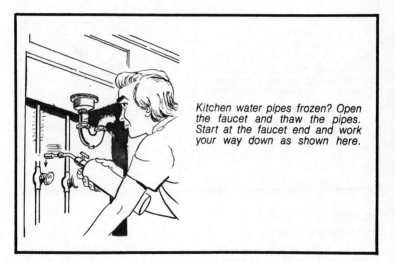

Kitchen water pipes frozen? Open the faucet and thaw the pipes. Start at the faucet end and work your way down as shown here.

Thawing Frozen Pipes

This is a common problem during severe winter weather. But doing it the wrong way can cause pipes to burst. Never, never start heating at the middle of a pipe run. Start at a faucet end. Open up the faucet and keep working your way toward the water supply. You can tell if everything is in apple pie order by the water which will start dripping out of the open faucet. Pipes underneath a sink which backs up to an outside wall are the ones to watch out for during freezing weather. A good idea is to leave the cabinet doors of the sink open so that some of the room heat will protect the pipes.

Use the torch to melt ice from the walkway. Melt it just enough so that a stiff broom can sweep it away; works faster than salt.

Clearing Steps of Ice

Rock salt and chemicals induce an action in concrete and bricks that is called "spalling." Chopping away the ice only tends to aggravate this situation. A better idea is to use the wide-flame attachment of the **propane gas** torch to partially melt the ice. Apply just enough heat for the ice to loosen its grip on the concrete; then use a stiff bristle broom to brush the ice away—do not chop it away.

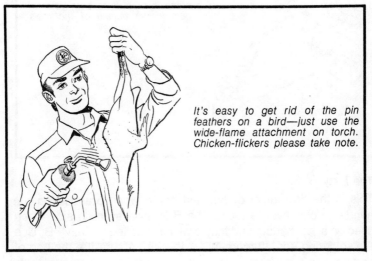

It's easy to get rid of the pin feathers on a bird—just use the wide-flame attachment on torch. Chicken-flickers please take note.

A Fowl Deed

We know of an outdoorsman who uses his propane gas torch to prepare his quota of ducks for the table. After dry-plucking the duck, he lights the torch to give the bird an over-all singeing to remove the pin feathers. A vast improvement over the way we recall our mother doing it over a blazing newspaper!

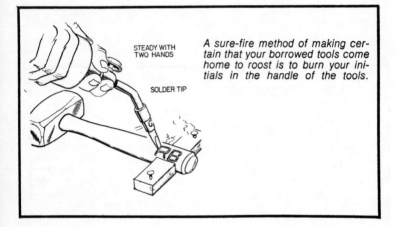

STEADY WITH TWO HANDS

SOLDER TIP

A sure-fire method of making certain that your borrowed tools come home to roost is to burn your initials in the handle of the tools.

The Lazy B

This is the first initial of our last name when branded on the handle of our favorite tools. The B is always on its side when used on a tool handle and that is why it is called a "lazy" B. Use the soldering iron attachment for this job. A sure-fire method of getting your tools back—after all, who would want to own tools with someone else's initials on them? Yes, some people would!

A neighbor of ours has a most elaborate collection of hand and power tools and he has solved the tool-lending problem rather nicely. He has a second set of tools, somewhat inferior in quality to his personal tools—and it is these tools that he lends out—and he has the reputation of being a good neighbor!

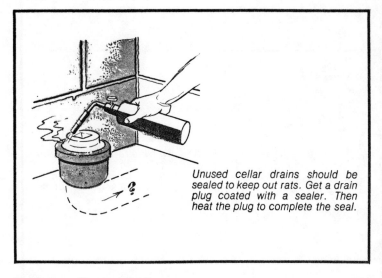

Unused cellar drains should be sealed to keep out rats. Get a drain plug coated with a sealer. Then heat the plug to complete the seal.

Sealing an Unused Drain

It is always a good idea to seal unused drains in the basement. Sometimes a drain may have been installed in a basement to take care of a water problem—which since disappeared—or possibly the builder installed a drain which leads to nowhere. Such a drain should always be sealed. Get a drain plug, available from hardware stores, that is coated with a sealer. Insert the plug, then apply heat to effect the seal. A friend of ours once walked into his basement darkroom and was confronted by a belligerent rat eyeing him from the top of a paper safe. He left the darkroom in a bit of a hurry, closed the doors and called the Health Department. They advised setting traps. So for 3 days he gingerly went down to his basement to inspect the traps he had set. No luck. Nary a nibble. This must have been a smart and sophisticated city rat with a disdain for food that was not of gourmet quality. More frantic calls to the Health Department. A man was finally dispatched to his home with a small caliber rifle —and that is how this rat was dispatched. A harrowing experience for his wife, too—so seal the traps; the Health Department conjectured that that was where the rat had come from.

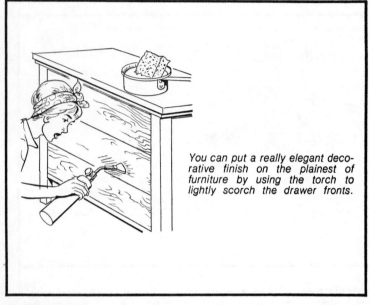

You can put a really elegant decorative finish on the plainest of furniture by using the torch to lightly scorch the drawer fronts.

Raising Grain

No, not wheat, corn or barley but the grain on unfinished furniture prior to finishing. Go over the entire piece of furniture with a well-dampened rag, wait a few minutes until the grain starts to rise. You can feel it by brushing your palm along the surface of the wood. Then use the propane gas torch with its wide-flame attachment. Keep the torch moving rapidly, or else you will wind up with a charred finish. This is somewhat on the order of singeing a fowl.

Whether your name is Clark, Carter or Katzenjammer, you can make an elegant home name plate by burning your name in a wood plank.

Custom Name Plate

Use the soldering tip of the propane gas torch to burn your name into a plank of wood. Best to first spell out your name in chalk, so you won't run off the end of the board. Wait until the tip has become red hot and apply your penmanship to the board. When finished, brush out any ash with a whisk broom and go over the entire surface of the board with a couple of coats of exterior varnish. Don't use shellac; the elements will turn it white.

Instead of using smelly liquids to start your outdoor barbecue, use the torch to get the charcoal burning—faster, cleaner and safer.

First Aid for the Barbecue

The party is all set, the hungry guests have arrived and there you are struggling over a balky bed of charcoal that seems unable to do its job. Never mind the comments and advice of assorted small fry, just haul out the propane gas torch from your workshop and direct its flame to the charcoal. In less time than it takes to call out, "Come and get it," the charcoal will be glowing —ready for steaks and hot dogs.

Remodeling the bathroom? Use the torch to heat the cement holding the old fixtures in place while gently prying with a putty knife.

Removing Epoxy

Contrary to what many people believe, cured and hardened epoxy can be removed. It all boils down to the proper application of heat. Use the pencil point heat attachment on the torch and apply it to the epoxied area. As soon as the epoxy gets hot, it will start to bubble and can be scraped or pried off. Work fast, don't allow the epoxy to cool; however, if it does, apply more heat. Tell your skeptical friends that epoxy *can* be removed.

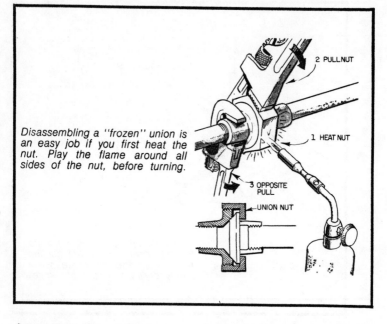

Disassembling a "frozen" union is an easy job if you first heat the nut. Play the flame around all sides of the nut, before turning.

2 PULL NUT

1 HEAT NUT

3 OPPOSITE PULL

UNION NUT

Loosening Pipe Joints

The pipe joint which is usually the most difficult to loosen—without damage—is a pipe union. This is so because there is a threaded joint at either side of the union and, of course, the union itself consists of male and female pipe threads. Unnecessary force on the nut of the union may instead loosen the pipe joints at either side of the union—after all, you are not trying to loosen or undo these joints—it is the union that you are trying to uncouple.

Apply the flame of the torch to the complete circumference of the union nut. Be especially careful if the joints at either side of the union are soldered joints. If they are, protect them with wet rags. The heat of the torch will cause the nut to expand slightly, just enough to break away from the male part of the thread.

Caution: It is always a good idea to use a second wrench to support and hold the pipe while trying to loosen the union nut. The second wrench will prevent the pipe from twisting as you apply force to the union.

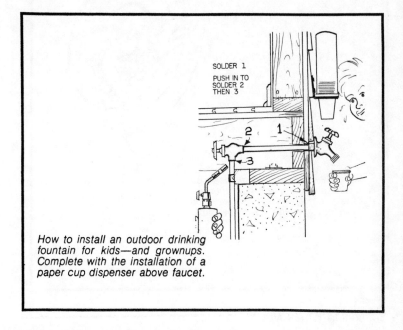

SOLDER 1

PUSH IN TO
SOLDER 2
THEN 3

How to install an outdoor drinking fountain for kids—and grownups. Complete with the installation of a paper cup dispenser above faucet.

An Outdoor Drinking Fountain

"I want a drink of water!" A commendable ambition, as my aunt used to say when I was a youngster. On warm summer days youngsters playing outdoors may come frequently to the kitchen sink for a drink of water. These visits can become a nuisance. Stop them by providing paper cups outdoors near a hose faucet.

Plastic dispensers for wall mounting and paper cup refills are available from supermarkets. You may already have one in your bathroom or kitchen.

If possible, mount the dispenser outdoors where it will be sheltered from rain. Keep a plastic wastebasket or bag nearby for disposal of used cups.

Consider installing a new hose faucet if you have none in a spot to do double duty as a source of drinking water. Run a copper-tubing cold water line to the patio—or adjacent to the spot in the yard where you relax or the kids play. Use a Bernz-Omatic torch to sweat-solder the tubing joints. Be sure to choose a frost-free faucet for any new installation.

Too much muscle can shear off wheel studs; better to heat nuts with the torch, apply light oil and then use the lug wrench to remove nuts.

Loosening Wheel Nuts

When it comes time to remove the snow tires—or install them—chemicals and salts used on the roads have a tendency to "freeze" the nuts to the wheel studs. Don't use brute force to loosen the nuts. You may either shear off the studs or chew up the nuts. A better way is to apply a few drops of light oil to the exposed threads of the studs, if they are visible, or around the nuts if the stud threads are hidden. Then use the torch to lightly heat the nuts and the studs. The heat of the flame will thin the oil and drive it into the threaded area. The heat will also break the seal between the nut and the stud. Apply oil and heat to all the nuts and studs. Now use the wheel wrench. Chances are that the problem now no longer exists. You can remove the nuts.

Those pesky caterpillars, weevils and insects of similar ilk can be destroyed by the torch. The same applies to vermin in iron bedsteads.

The Torch as a Vermin and Insect Killer

Those hairy caterpillar nests which seem to arrive in cycles (no doubt to the configuration of the planets) can be quickly destroyed with a pass or two of the propane gas torch. You may need a ladder to get at some of them but it is a fast and easy way of destroying these pests—and much faster and more satisfying than using a spray gun.

We know of a motel-keeper who uses a torch on a regular basis to keep his beds free of bedbugs. Formerly he squirted a few drops of gasoline into all the joints of beds that had a steel framework. Then he lit the gasoline. Risky. Now all he does is make a few passes over any suspected areas with the torch—and the job is done. Safely.

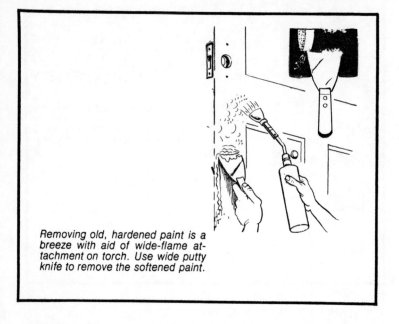

Removing old, hardened paint is a breeze with aid of wide-flame attachment on torch. Use wide putty knife to remove the softened paint.

Removing Paint

Of course, the propane gas torch is an ever-so-handy device when it comes to removing paint. No chemical paint remover can remove a half-dozen layers of paint that have been applied to an exterior door and that has been beaten into a rock-like mass by the sun.

Use the wide-flame attachment on the torch. As the paint starts to soften, scrape it off with a wide putty knife. Stubborn paint areas that seem to defy the putty knife can be removed with the aid of coarse steel wool.

Tip: Round the edges of the putty knife and you will remove any chances of gouging the wood during the scraping operation. Use a file or a grinder to round the corners.

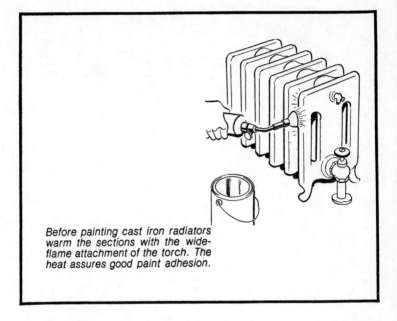

Before painting cast iron radiators warm the sections with the wide-flame attachment of the torch. The heat assures good paint adhesion.

Painting Radiators

Paint is less apt to flake off—and will make a better bond—if the painting is done while the radiator is warm. Warm, but not hot to the touch. Then, after the paint has been applied and the radiator really starts to get hot, it will bake the paint in place.

But such painting chores are usually left to the spring and summer time when radiators are cold. And who wants to turn on the furnace during warm weather? Just for the sake of heating a couple of radiators? Well, simplify your work—and work in comfort—by applying the heat of the propane gas torch to each section of the radiator just prior to painting. Use the wide-flame attachment. Brush the flame up and down, just long enough to warm the radiator.

Index

Robert Brightman was Home and Shop Editor of *Mechanix Illustrated* magazine for 24 years where he assigned, wrote and edited articles on all facets of home maintenance, repairs, wiring, alterations and furniture making. He also conducted a question-and-answer column on home problems called "The House Doctor." At present he writes for *Popular Mechanics, Better Homes and Gardens, McCalls,* and kindred publications. He is also responsible for the current *Reader's Digest Complete Do-it-Yourself Manual* and for many Stanley Tools instruction booklets. His latest book is the *Home Owner's Handbook of Carpentry and Woodworking.* In addition, he has edited 14 books on electricity, masonry, concrete, plumbing, home repairs, heating, astronomy and photography. Bob Brightman has appeared on more than 100 TV and radio shows slanted to the homeowner. He is a homeowner who spends what little spare time he has keeping it shipshape.

Henry Clark is a long-time illustrator of how-to articles. He knows what he is drawing as he has been building homes inside and out for the past 25 years.